Performance-Based Curriculum for Science

From Knowing to Showing

Helen L. Burz
Kit Marshall

Performance-Based Curriculum for Language Arts
Performance-Based Curriculum for Mathematics
Performance-Based Curriculum for Science
Performance-Based Curriculum for Social Studies

Performance-Based Curriculum for Science

From Knowing to Showing

Helen L. Burz
Kit Marshall

CORWIN PRESS, INC.
A Sage Publications Company
Thousand Oaks, California

Copyright © 1997 by Corwin Press, Inc.

All rights reserved. Use of the blank templates in the Appendix is authorized for local schools and noncommercial entities only. Except for that usage, no other part of this book may be reproduced or utilized in any form or by any means, electronic or mechanical, including photocopying, recording, or by any information storage and retrieval system, without permission in writing from the publisher.

For information address:

Corwin Press, Inc.
A Sage Publications Company
2455 Teller Road
Thousand Oaks, California 91320
e-mail: order@corwin.sagepub.com

SAGE Publications Ltd.
6 Bonhill Street
London EC2A 4PU
United Kingdom

SAGE Publications India Pvt. Ltd.
M-32 Market
Greater Kailash I
New Delhi 110 048 India

Printed in the United States of America

Library of Congress Cataloging-in-Publication Data

Burz, Helen L.
    Performance-based curriculum for science: from knowing to showing /
      Helen L. Burz and Kit Marshall.
       p.   cm. — (From knowing to showing)
    Includes bibliographical references.
    ISBN 0-8039-6506-0 (cloth : alk. paper) — ISBN 0-8039-6507-9 (pbk. : alk. paper)
    1. Science — Study and teaching.   2. Curriculum planning.
    3. Competency-based education.
  I. Marshall, Kit.   II. Title.   III. Series
Q181.B967 1997
507'.1'2 — dc20   97-4946

This book is printed on acid-free paper.

   98 99 00 01 02   10 9 8 7 6 5 4 3 2

Corwin Press Production Editor: S. Marlene Head
Typesetting: Birmingham Letter & Graphic Services
Cover Designer: Marcia R. Finlayson

TABLE OF CONTENTS

PREFACE vi

ABOUT THE AUTHORS viii

INTRODUCTION 1
    IMPORTANT QUESTIONS AND ANSWERS ABOUT *PERFORMANCE-BASED CURRICULUM FOR SCIENCE* 1
    PERFORMANCE-BASED LEARNING ACTIONS 5
    THE PERFORMANCE-BASED LEARNING ACTIONS WHEEL 6

1. CONTENT/CONCEPT STANDARDS FOR SCIENCE 11
    WHY IS SCIENCE IMPORTANT? 11
    VISION 11
    PROGRAM GOALS 11
    PROCESS SKILLS 12
    CONTENT STRANDS 12
    CONTENT/CONCEPT STANDARDS 13
    PERFORMANCE BENCHMARK FORMAT 13
    LIFE SCIENCE–LIVING THINGS 15
    LIFE SCIENCE–ECOSYSTEMS 21
    PHYSICAL SCIENCE–MATTER 27
    PHYSICAL SCIENCE–ENERGY 33
    EARTH SCIENCE–METEOROLOGY 39
    EARTH SCIENCE–GEOLOGY 44
    EARTH SCIENCE–SPACE SCIENCE 50

2. TECHNOLOGY CONNECTIONS 57
    SUMMARY 57
    SKILLS AND ABILITIES 59
    PHYSICAL SCIENCE–MATTER: TECHNOLOGY CONNECTIONS 64
    EARTH SCIENCE–GEOLOGY: TECHNOLOGY CONNECTIONS 68

3. PERFORMANCE DESIGNERS 73
    PERFORMANCE DESIGNER FORMAT 73
    EXAMPLES OF LEARNING ACTIONS 78
    SAMPLE PERFORMANCE DESIGNERS 79

APPENDIX: BLANK TEMPLATES 87

BIBLIOGRAPHY 95

PREFACE

Traditionally, textbooks and curriculum guides have reflected a focus on content coverage. Districts, schools, and educational systems have looked to publishers to define, at least in general terms, *what* should be taught and the order in which it should be taught. The result has been to place an emphasis on what students need to *know*, often with little direction regarding the role of relevance and meaning for the learning.

The impact of technology on society and a scan of future trends clearly deliver the message that just teaching information and "covering the book" are no longer sufficient as a focus for instructional systems. Instead, instruction must go beyond the content taught and actively engage learners in demonstrating how they can select, interpret, use, and share selected information. Educators are quick to accept this shift but are faced with a real need for models that depict ways this might occur.

Performance-Based Curriculum for Science provides a unique model for taking instruction from the traditional focus on content to a student-centered focus that aligns selected content with quality and context.

Because of the focus on content related to a particular content discipline, textbooks and curriculum frameworks and guides have had a strong influence on *how* content is taught. The result, often, has been to teach facts and basic functional skills without using a meaningful, learner-centered approach. There has been no purpose in mind beyond having students know certain information and skills. These previous frameworks and guides have also separated curriculum from instruction and assessment. *Performance-Based Curriculum for Science* offers a new organization and alignment of curriculum, instruction, and assessment around practical classroom application and does it in a way that readily allows teachers to use it.

Although not intended to be a complete daily curriculum guide, *Performance-Based Curriculum for Science* provides a planning framework that includes numerous examples of performance-based science set in real-life contexts. The numerous performance benchmarks, at Grades 3, 5, 8, and 12, and strands can be used directly or as guides for customizing instruction toward relevant and meaningful application of important knowledge around critical scientific concepts. *Performance-Based Curriculum for Science* can be used to guide the development of science curriculum throughout a family of schools or by individual teachers within one classroom or by an instructional team.

The book is divided into four major sections:
1. Introduction to *Performance-Based Curriculum for Science*
2. Content/Concept Standards for Science and Performance Benchmarks for 3rd, 5th, 8th, and 12th Grades
3. Technology Connections
4. Performance Designers

The Introduction is organized around a friendly question-and-answer format. This section is central to the planning framework and provides the rationale and organizational structure for the book. The Introduction also contains a discussion of performance-based learning actions.

The Content/Concept Standards for Science represent the best thinking of current national experts and provide the substance for each performance benchmark. These standards are organized by major strands within the discipline. Performance Benchmarks included in this section represent descriptions of what could be expected from a student who has a high degree of understanding of a content standard in a high-quality performance. For example, the student might be asked to solve a real-life problem or develop alternative solutions to an issue or question that requires a solid understanding of the content/concept standard at one of four developmental levels.

Technology Connections provide guidance for the application of technology in some portion of a performance benchmark. These strategies are appropriate for students who are accessing, producing, and disseminating information through technology.

The last section, Performance Designers, provides an analysis of the performance designer, which is a planning tool for teachers. It requires a focus on the key elements of content, competence, context, and quality criteria.

At the end of the book, design templates and reproducible masters (see Appendix: Blank Templates) provide practical tools that can be used to customize and create classroom instructional material that will empower teachers and students to be successful in "showing what they know."

ABOUT THE AUTHORS

HELEN L. BURZ

Helen L. Burz is a doctoral candidate at Oakland University in Rochester, Michigan, where she received her master of arts degree in teaching. She received her bachelor of science in education from Kent State University. Helen has taught at the preschool, elementary school, and college levels. She has also worked as a principal at the elementary and middle school levels. As an innovative leader in curriculum design and instructional delivery systems, she has led her schools to numerous state and national awards and recognition and was selected as Administrator of the Year in Michigan.

She has addressed integrated curriculum and interdisciplinary instruction for the Association for Supervision and Curriculum Development's (ASCD's) Professional Development Institute since 1985. Currently, she works as an educational consultant across North America, speaking and conducting training for future-focused, performance-based curriculum, instruction, and assessment.

KIT MARSHALL

Kit Marshall earned her Ph.D. at Stanford University in educational leadership in 1983 and her master's and BA at Sacramento State University in 1968. After teaching across all levels, developing state and national dissemination grants in innovative educational design, and site-level administration, she pursued further studies in organizational development and technology. She has received numerous awards for her work in restructuring curriculum, instruction, and assessment. Her book, *Teachers Helping Teachers*, published in 1985, was the first practical handbook for educators on team building and mentor teaching.

Currently living in California, Marshall is an international speaker and trainer in future-focused, performance-based curriculum, instruction, and assessment. She is CEO of Action Learning Systems, an educational restructuring company and President of The Learning Edge, a World Wide Web (WWW) site dedicated to networking restructuring schools and communities throughout North America.

INTRODUCTION

Authentic *performance-based education* asks students to take their learning far beyond knowledge and basic skills. A *performance orientation* teaches students to be accountable for knowing what they are learning and why it is important and asks them to apply their knowledge in an observable and measurable *learning performance.*

This shift "from knowing to showing" means that everything we do—instruction, curriculum, assessment, evaluation, and reporting—will ultimately be focused on and organized around these learning performances.

Educators, parents, business and industry leaders, and community members throughout North America are coming to agree that students should be demonstrating what they are learning in observable and meaningful ways. However, we have all been to school. Generally, our collective experience of what school *is* has been very different from what we believe schools need to *become.* If we are to succeed in the difficult shift from content coverage to performance-based education, we will need to have new strategies for defining and organizing what we do around *significant learning performances.*

Performance-Based Curriculum for Science has been developed to provide the tools and the structure for a logical, incremental transition to performance-based education. *Performance-Based Curriculum for Science* is not intended to be a comprehensive curriculum; it is a curriculum framework. The various components of the framework provide structure and a focus that rigorously organizes *content* around *standards* and *performance* around *learning actions.*

IMPORTANT QUESTIONS AND ANSWERS ABOUT *PERFORMANCE-BASED CURRICULUM FOR SCIENCE*

Content/Concept Standards

Where do the content/concept standards come from for this framework?

This framework represents the best thinking of current national experts in the discipline of science. Although there is no official national standard for content areas, the National Science Teachers Association and the American Association for the Advancement of Science's Project 2061 and its *Benchmarks for Science Literacy* have demonstrated strong national leadership and influence that could form the instructional focus in a K–12 science program. These recommendations have been used to form the content/concept foundation of this framework and are identified as content/concept standards.

How are the content/concept standards organized within this framework?

The content/concept standards for science are organized by major strands and substrands within the domains of the natural sciences. These strands are listed and described in Chapter 1. For example, the strand of life science contains two substrands: living things and ecosystems.

How do I know which content/concept standards to focus on with MY students?

What students should know by the end of four levels, specified as Grades 3, 5, 8, and 12, is described at the beginning of each content Strand section in Chapter 1. These levels are identified to highlight the specific developmental stages the learner moves through in school. A 1st-grade teacher should teach to the development of the concepts identified at Grade 3. A 6th-grade teacher should use the 5th-grade and 8th-grade content/concepts to guide instruction. A 9th-grade or 10th-grade teacher should use the 8th-grade contents as a guide and teach to the 12th-grade content/concepts.

These identified standards provide the content/concept focus for the performance benchmarks within the discipline and within the four developmental levels. Each major strand is identified by a set of content/concepts standards and is followed by four performance benchmark pages: one at each of the four levels—3rd, 5th, 8th, and 12th grade.

Performance Benchmarks

What is a performance benchmark?

In *Performance-Based Curriculum for Science*, a performance benchmark is a representative description of what could be expected from a student who has a high degree of understanding of a content standard and can use that content standard in a high-quality performance. For example, the student might be asked to solve a real-life problem or develop alternative solutions to an issue or question that requires a solid understanding of the content/concept standard. If the students don't have the knowledge, they will not do well in the benchmark.

Each performance benchmark is designed to target a particular developmental level identified as 3rd, 5th, 8th, and 12th grades. Many students will be able to perform at a higher level, and some will perform at a lower level at any given point. Where a student is in the benchmarking process will determine where he or she is in the continuous learning process so characteristic of performance-based education.

What are the components of a performance benchmark?

Each performance benchmark has:

1. A **Key Organizing Question** that provides an initial focus for the performance benchmark and the content/concept standard addressed in the performance benchmark.

2. Performance-based **Key Competences (Learning Actions)** that specify what students need to do with what they know in the performance benchmark (refer to Figure 1.1, The Learning Actions Wheel, on page 6).

3. **Key Concepts and Content** from the discipline that define what students need to know in the performance benchmark.
4. **Two Performance Tasks**, or prompts, that provide the purpose, focus, and authenticity to the performance benchmarks. Having two tasks allows a teacher to ask for a group or individual performance, or even to ask for a repeat performance.
5. **Quality Criteria** or **"Look fors"** that precisely describe what a student would do to perform at a high-quality level on that performance benchmark. This component serves as the focus for the evaluation process. How well students can demonstrate what is described in the quality criteria informs the evaluator about continuous improvement planning goals for a student. The profile that results from an entire classroom's performance benchmark informs the teacher regarding next steps in the teaching-learning process.

How do I use the performance benchmarks to inform and guide ongoing instruction and assessment?

The performance benchmarks will:

- Organize *what* you teach around a clear set of content/concept standards for a particular discipline
- Organize *how* you teach by focusing your planning on the learning actions that you will teach and assess directly during daily instruction
- Provide you with specific targets for your instruction—you will teach "toward" the performance benchmarks
- Focus your students on what they will need to demonstrate in a formal evaluation of their learning
- Communicate to parents that there is a clear and rigorous academic focus to authentic performance-based education

The **performance benchmarks** are primarily for evaluation of learning, *after* the learning has occurred. The **performance designer**, on the other hand, provides the focus for quality continuous improvement *during* the ongoing daily instructional process.

Technology Connections

How about a technology connection for Performance-Based Curriculum for Science?

A number of performance benchmarks in *Performance-Based Curriculum for Science* have a companion application that uses technology in some portion of the performance. If students are currently accessing, producing, and disseminating using technology, you will want to use the strategies found in this section. These technology connections also serve as examples for teachers who are just moving toward the use of technology in their classrooms.

Computer Icon

If there is a computer icon on the performance benchmark page, you can refer to the companion page that will extend the performance benchmark to involve technology.

Performance Designers

What is a performance designer?

A performance designer is an organizer that is used to plan for ongoing performance-based instruction and assessment. The performance designer in *Performance-Based Curriculum for Science* uses the learning actions and connects them to content, context, and criteria. The power of these learning actions becomes apparent when students begin to recognize and improve their competence with each new learning performance.

How is the performance designer used?

The performance designer can be used to organize student performances in any discipline and with students at all developmental levels and in all grades.

The sample performance designers provided can be used just as they are or can serve as a starting point for new designs.

You are invited to copy the blank performance designers in the Appendix for your own classroom use, or you may want to create a new performance designer that fits your style of planning and thinking.

How can I design performances for my students?

Performances can be designed by following the steps provided in Chapter 3 on performance designers.

PERFORMANCE-BASED LEARNING ACTIONS

Learning actions organize what the students will *do* with what they *know* in each performance benchmark. Performance-based learning actions are based on four important beliefs:

1. *Learning is a quality continuous improvement process.*	Students improve their performance with any learning when they have multiple opportunities to apply what they know in a variety of settings over time. As students become familiar with and adept at using certain key learning actions, the quality of each subsequent performance will improve. Students will be *learning how to learn*.

2. *Certain learning actions, or competences, apply to the teaching/learning environment regardless of the age of the learner or the content being taught.*	The five performance-based learning actions coupled with continuous assessment and evaluation are applicable to all ages and in all content areas. The current level of competence with these learning actions will vary from student to student. There will be a considerable range of competence with these learning actions even within a single classroom or grade level. The focus of improvement is on comparison to a learner's last best effort, not comparison of students to one another and not on the content alone. Performance-based teaching and learning will focus on what students can *do* with what they *know*.

3. *Successful people are able to apply certain key actions to every learning challenge. These actions have similar characteristics regardless of the challenge.*	When students learn, apply, and continuously improve in the learning actions, they are practicing for life after they leave school. Schools must allow students to practice for the challenge, choice, and responsibility for results that they will encounter after "life in school" is over. The more competent students are with a range of these learning actions, the more successful they will be in dealing with the diverse issues, problems, and opportunities that await them.

4. *The problem with the future is that it is not what it used to be.*	Today's informational and technological challenges mean that schools must restructure themselves around a different set of assumptions about what students need to *know* and be able to *do*. Many educators and parents are reaching the conclusion that much of the information we ask students to remember and many of the skills we ask them to practice may no longer be appropriate or useful by the time they leave school. At this point, we ask the question, "If covering content is not enough anymore, what *should* schools be focusing on?" We believe the answer is "The learning actions."

THE PERFORMANCE-BASED LEARNING ACTIONS WHEEL

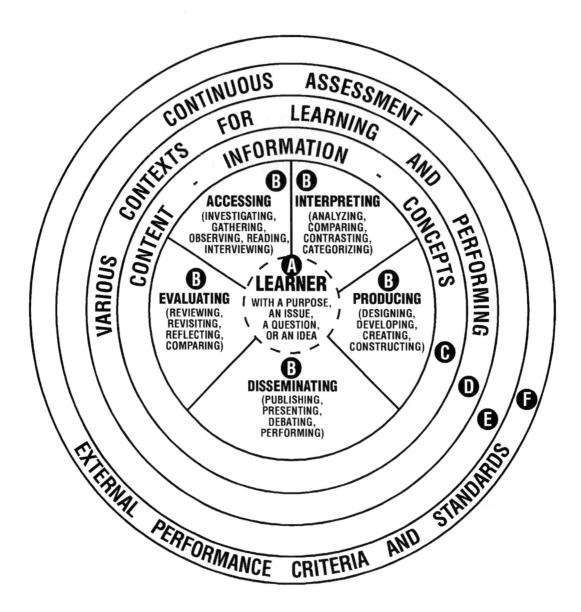

FIGURE 1.1 THE LEARNING ACTIONS WHEEL

Ⓐ The Learner

The learning actions are learner centered and brain based. At the center of the Wheel in Figure 1.1 is the learner with a stimulus for learning. That stimulus may be an issue, idea, or question that may have been suggested to the teacher by the content standards, or it may be something of particular interest to the learner. The learner is in the center because no matter how important we think the content is, it is inert until we add action to it. Everything "revolves" around the developmental levels, the motivation, and the engagement of the learner in the learning actions.

❸ The Five Major Learning Actions of a Performance

The learning actions include five major stages that learners will move through during any performance process. Let's look at the meaning and importance of each.

Accessing

What do I need to know?

How can I find out?

A performance begins with an issue, a problem, or an interesting "lead." The learner accesses the information he or she needs to have in order to successfully perform. This information can come from a variety of experiences—but it must come from somewhere. Traditionally, information has come solely from the teacher or the next chapter in a textbook. In today's information-based environment, students must be adept, self-directed learners, determining what is needed and having a wide range of competences for accessing critical information and resources. Learners may investigate, gather, observe, read, and interview, to name a few actions. Whatever actions they engage in to find out what they need to know, *relevant* information must be accessed if the performance is to be as powerful as possible. Accessing is an important first step to a performance and a critical component of success as a learner in any role in or outside school.

Interpreting

What does all of this mean?

So what?

Critical reasoning, problem finding and solving, decision making, and other similar mental processes are what we must do as a part of any important learning that we intend to use in some way. Here, we must make sense of the information we have accessed and decide which information to keep, expand on, or ignore. This component of a performance asks us to analyze, compare, contrast, and categorize—to somehow meaningfully organize the information to represent what we think it all means. This component is critical to a performance process. It clearly determines the level of sophistication and competence with which we can deal with the amount of information constantly vying for our attention and time both in and outside a formal learning situation.

Producing

How can I show what I know?

What impact am I seeking?

Who is my audience?

The producing component of a performance is when we translate what we have learned into a useful representation of our learning. What gets produced represents a learner's competence with design, development, creation, and construction—something tangible that pulls the learning together in some form. This component is the acid test of a learner's competence as a quality producer, a critical role for working and living in the 21st century. In life after school, what we produce usually has a focus, an audience in mind. A powerful performance will always have a clear purpose in mind, a reason for the performance, and an impact that is desired as a result of the performance.

Disseminating

What is the best way to communicate what I have produced?

How can I impact this audience?

How will I know that what I've produced has had an impact?

The fourth learning action of the performance process is disseminating. At this point, we are asking the learners to communicate what they have learned and produced to someone, either directly or indirectly. This is also where the value of an authentic context, someone to be an involved and interested audience, is so apparent. Only in school does there seem to be a lack of attention paid to such a critical motivation for learning and demonstrating. This is truly the point at which learners are dealing with the challenges of a performance setting. Students may publish, present, debate, or perform in a variety of fine and dramatic arts activities, to name a few possibilities. Service learning projects, community performances, and a variety of related school celebrations of learning are all ways for the learning to hold value that may not be inherently present in the simple existence of content standards.

Evaluating

How well did I do?

Where will I focus my plans for improvement?

The evaluation component represents the culmination of one performance and perhaps the launching of another performance cycle. It is the point at which a judgment is made and plans are developed for improvement next time. The quality criteria (in the performance benchmarks) for all the learning actions are the guides for these evaluations. The performance benchmarks in this framework represent the personal evaluation component of ongoing learning and performing on a day-to-day basis.

❸ Content–Information–Concepts

The learning actions are applied to the content–information–concepts identified by the educational systems as being essential. Addressing information that is organized around major concepts allows the learner to work with a much broader chunk of information, and thus the learner is afforded the opportunity for making more connections and linkages.

In this text the selected information has been aligned with the content/concepts recommended by the National Science Teachers Association and the American Association for the Advancement of Science's Project 2061 and its *Benchmarks for Science Literacy*.

❹ Various Contexts for Learning and Performing

A context for learning refers to the setting in which the learning occurs, or the audience or recipient of the fruit of the learning or the situation—any of which create a reason, purpose, or focus for the learning.

Traditionally, the context for learning for students has been alone in a chair at a desk in a classroom. However, the context can be a river or stream that runs through the community. Students working in groups with engineers from a local plant can be engaged in collecting specimens and conducting experiments from the water to determine effects of manufacturing on the water's purity, so they can submit a report to the company or the Environmental Protection Agency.

E Continuous Assessment

The continuous assessment circle of the learning actions wheel represents the continuous improvement process that is imbedded throughout each of the other components. An authentic learning community will engage in a supportive improvement process that is less competitive than it is collaborative and cooperative. To *assess* originally meant "to sit beside." During key points in each component of the performance process, students will reflect upon their own work and the work of others. The role of the teacher in this process is to ask questions that guide the student's self-assessment and provide specific feedback to the learner about what is being observed. The conditions we create for this reflective assessment on a daily basis will determine the ultimate success students will have with the performance benchmarks.

F External Performance Criteria and Standards

The outermost circle represents the system's standards and scoring or grading procedures and patterns.

Remember, there are four critical components of a performance. The learning actions represent an organizing tool for a performance. They describe the components of the performance process. The learning actions also represent quality work according to identified criteria. By themselves, the learning actions are of little use. You have to *know* something to *do* something with it. In *Performance-Based Curriculum for Science*, each performance benchmark combines all four components of a performance:

1) Content–information–concepts
2) Competence: learning actions performed by the learner
3) Contexts that create a reason and a focus for the performance
4) Criteria that define a quality performance

1
CONTENT/CONCEPT STANDARDS FOR SCIENCE

WHY IS SCIENCE IMPORTANT?

This section is organized through the use of content strands, processes, and concepts as identified by the American Association for the Advancement of Science's Project 2061: Science for All Americans and its *Benchmarks for Science Literacy*, the National Science Teachers Association, and Science for All Students from the Florida Department of Education. This content is not exhaustive but instead represents a shift from too much content treated superficially, to fewer topics with more in-depth coverage. It recommends concepts that all students should know and provides examples of performance benchmarks, which are classroom events that require students to apply the designated key concepts and content.

The explosive growth of scientific knowledge and continual developments in technology are transforming society. Homes, schools, and places of work are much different now than they were a few years ago. Consequently, science education should be designed for lifelong learning in a world shaped by science and technology. Science is a basic part of every student's education.

VISION

The science program will prepare each student to become a knowledgeable citizen who is able to make informed decisions in a technological society. All students will have the opportunity to study science in an interesting and worthwhile way that will open their minds to new outlooks and equip them with the intellectual skills that will guide their learning for the rest of their lives. Working in supportive, cooperative, nurturing, and respectful environments in which instruction and curriculum reflect identified standards, students will engage in activities that enable them to understand the connections between science and technology and the practical applications of this knowledge to their real-life experiences. Students will observe, analyze, construct models, ask questions, combine logic with imagination, and make rational systematic predictions and explanations to better understand the natural world.

PROGRAM GOALS

Science literacy for students is defined in seven programmatic goals. These goals emphasize the use of conceptual themes and scientific methods of reasoning and thinking to provide valuable tools that can be used for lifelong learning. The goals are to

1. Develop positive approaches toward learning science and develop ways of thinking scientifically
2. Acquire a broad base of scientific knowledge that will enable students to understand and interpret natural world phenomena
3. Develop a knowledge and understanding of the history of science from a multicultural perspective
4. Recognize and understand the application of science to personal life and living
5. Develop and utilize integrated thinking skills in daily problem-solving situations
6. Develop an understanding of the interrelationships and scientific, technological, and societal issues that affect the decision making of a scientifically literate populace
7. Apply the processes of science in scientific investigations through hands-on experiences

PROCESS SKILLS

In science, a strong emphasis is placed on developing students' ability to employ the processes of science and to learn about the natural phenomena around them. These same processes are used by scientists in pursuing their own investigations. Demonstrations of the process skills enable students to "do science." They permit "hands-on" or "performance-based" science.

Students develop facility with the basic processes in early grades and move on to the more complex integrated process in Grades 4 through 8. The laboratory science courses in senior high school synthesize the use of
all of the processes into a set of investigative skills that can be used in the lifelong pursuit of solutions to problems.

The basic processes include:

> observing, describing, ordering, inferring, organizing space/time relationships, comparing, measuring, communicating, classifying, predicting, and formulating questions.

The integrated processes include:

> formulating hypotheses, controlling variables, collecting and interpreting data, defining operationally, experimenting, concluding, and recommending.

CONTENT STRANDS

The science program contains the content and processes of the three major domains of scientific study—life science, physical science, and earth/space science. Additionally, it contains the technologies or processes of science.

The science content defines the scientific terms, facts, laws, principles, concepts, theories, applications, tools, and instruments that every student should understand and be able to use. Science content also defines the attitudes and abilities to apply scientific principles and reasoning that students should develop, as well as the nature of experiences that contribute to acquiring scientific understandings, attitudes, and abilities.

The performance benchmarks presented in this text are designed to address the following domains of the natural sciences:

Strands	Substrands
LIFE SCIENCE	LIVING THINGS
	ECOSYSTEMS
PHYSICAL SCIENCE	MATTER
	ENERGY
EARTH SCIENCE	METEOROLOGY
	GEOLOGY
	SPACE SCIENCE

CONTENT/CONCEPT STANDARDS

The content/concept standards identified and listed for the four developmental benchmarks are greatly influenced by the work done by the National Science Teachers Association and the American Association for the Advancement of Science's Project 2061 and its *Benchmarks for Science Literacy*.

The content presented in the performance benchmarks is not intended to be a complete, detailed list of all information students should know but rather represents essential ideas, concepts, and categories of science information and skills. Educators should consider these examples as a guide for their own selection process as they relate these ideas to locally identified curricula and expectations.

The examples included in this chapter address domains of the natural sciences. Each of the science strands and substrands is identified, briefly described, and then presented in terms of what students should know how to do by the end of Grades 3, 5, 8, and 12. Each strand will be introduced by a listing of the content/concept standards.

PERFORMANCE BENCHMARK FORMAT

The performance benchmarks are sample demonstrations designed with content, competence, context, and criteria that students should accomplish individually and collaboratively by the end of identified grade levels. For each of the substrands, there will follow four performance benchmarks. There will be one benchmark for each of the following developmental levels: 3rd, 5th, 8th, and 12th grade. Because these benchmarks represent different developmental levels, they should serve as guides for all teachers from kindergarten through 12th grade. The performance benchmarks are designed to represent a description of what could be expected from a student in a high-quality performance who has a high degree of understanding of the specific content/concept standard and has consistently experienced the learning actions.

The following template, along with descriptions, is offered as an advance organizer for the performance benchmarks that follow in the next section.

PERFORMANCE BENCHMARK FORMAT

A. SCIENCE STRAND AND SUBSTRAND AND STANDARD NUMBERS		G. TECHNOLOGY ICON	
B. KEY ORGANIZING QUESTION:			
C. KEY COMPETENCES	D. KEY CONCEPTS AND CONTENT	E. PERFORMANCE TASKS	
		PERFORMANCE TASK I:	
		PERFORMANCE TASK II:	
F. QUALITY CRITERIA: "LOOK FORS"			

A. Science Strand and Substrand and Standard Numbers

This section serves to identify the selected science strand and substrand and the specific standard numbers chosen from the content/concept standards pages that precede each set of benchmarks.

B. Key Organizing Question

Each performance benchmark addresses specific content information and is organized around a key organizing question. This question serves as a focusing point for the teacher during the performance. The teacher and student can use these questions to focus attention on the key concept/content and competences required in the performance task.

C. Key Competences

The key competences represent the major learning actions of accessing, interpreting, producing, disseminating, and evaluating. These major learning actions are discussed in detail on the preceding pages.

The actions identified are what the student will *do* with the key concepts and content in this benchmark performance. Those do's or learning actions engage students in demonstrations of competence in technical and social processes. Teachers must teach students how to operationalize these learning actions.

D. Key Concepts and Content

The information contained in this section identifies the major concepts that embrace the essential content and knowledge base that was taught and is now addressed in this performance benchmark. These concepts correspond to the standard numbers in Section A above.

E. Performance Tasks

Each performance task requires students to apply the designated content using specific learning actions they have been taught. This is done in a context or situation related to the key question. The performance tasks can be done individually or collaboratively. In either case, it is still the teacher's responsibility to look for the presence or absence of the quality criteria in action.

There are two performance tasks identified on each performance benchmark page to offer teachers a choice or serve as a parallel task for students. Both tasks correspond to the identified quality criteria.

F. Quality Criteria: "Look fors"

The quality criteria represent key actions that students are expected to demonstrate during the performance task. The criteria also guide the teachers and serve as "look fors" during the performance task. In other words, the teacher observes the students for these specific criteria.

These criteria embody the key competences or learning actions that students should have been taught in preparation for this performance task. Students demonstrate the learning actions in connection to the key concepts.

The criteria serve as a process rubric that guides the design of both instruction and assessment. They also serve as a signpost for the learners.

The criteria are identified following a "do + what" formula, which makes it easy to "look for" them.

G. Technology Icon

The presence of a technology icon at the top of a performance benchmark page means there is a corresponding example in the Technology Connections section. These examples indicate how technologies can assist students in carrying out the key competences required in the performance task.

LIFE SCIENCE – LIVING THINGS

Content/Concept Standards

People have long been curious about living things—how many different species there are, what they are like, where they live, how they relate to each other, and how they behave. Scientists seek to answer these questions and many more about the organisms that inhabit the earth. In particular, they try to develop the concepts, principles, and theories that enable people to understand the living environment better.

What students should know how to do by the end of Grade 3

Students should investigate the habitats of many different kinds of local plants and animals, including weeds, aquatic plants, insects, worms, and amphibians and some of the ways in which animals depend on plants and on each other. They should be able to

1. Compare and contrast the ways in which organisms in one habitat differ from those in another and examine how some of those differences are helpful to their survival
2. Identify and describe the characteristics of living things and demonstrate that they can be classified in different ways
3. Explore, identify, and evaluate how living things use their senses to learn and gather information
4. Predict the habitat of living things based upon an analysis of the observable characteristics of living things by exploring the relationship between structure, function, and environmental conditions

What students should know how to do by the end of Grade 5

Students should explore how various organisms satisfy their needs in the environments in which they are typically found. Students can examine the survival needs of different organisms and consider how the conditions in particular habitats can limit the kinds of living things that can survive. Studies of interactions among organisms within an environment should start with relationships that can be directly observed by students. The use of nature films and tapes could provide observations of diversity of life in different habitats. Students should be able to

1. Examine the effects of living things interacting with their environment
2. Investigate the relationships between the characteristics of parts of organisms or components of systems and the actions they perform
3. Explore, analyze, and explain the interdependence of plants and animals
4. Investigate the relationship between the environment, needs of the human body, and wellness
5. Research, identify, and explore the importance of diversity in the natural world and the behaviors and conditions necessary for survival

What students should know how to do by the end of Grade 8

As students build up a collection of information on organisms and habitats, they should be guided from specific examples of the interdependency of organisms to a more systematic view of the kinds of interactions that take place among the organisms. They should be able to

1. Investigate the cellular organization of living systems in terms of their structure, function, and energy requirements and investigate how the interactions of matter and energy determine the nature of their environment
2. Investigate and examine the relationships among energy availability, stability, and the evolution of living systems
3. Research rates and patterns of change in living systems and anticipate how some living systems may change in response to conditions in the environment

What students should know how to do by the end of Grade 12

All students should understand the structure of DNA and how it provides continuity from generation to generation. Mechanisms of change in DNA should be explored as well as the role of the environment in shaping the survival of these changes. In addition, students' understanding of the processes of the human organism should be expanded to include the functions of the basic systems, both in healthy and unhealthy persons. They should be able to

1. Investigate the relationships between the biochemical and anatomical processes in determining heritable characteristics
2. Research how new inherited characteristics can result from new combinations of genes as a result of genetic engineering and predict what implications this will have for humans and other living systems
3. Explore and investigate the nature of organisms in terms of the beneficial effects of some and the diseases that others may cause
4. Research and investigate the evolutionary processes that are supported by observable facts about life on earth in terms of diversity, similarity within that diversity, and the sequence of changes in fossils
5. Use the methods from microbiology to analyze and confirm the various theories from cell theory to disease theory and immunology
6. Investigate and explain the process of cellular reproduction beginning with the double helix theory of DNA structure and function
7. Research and investigate the postulates of natural selection
8. Explain what is known about the human body and explore what is needed to keep it well

Content/Concept Standards for Science 17

Science: **Performance**
Grade 3 **Benchmark**

LIFE SCIENCE: LIVING THINGS
CONTENT/CONCEPT STANDARD 3

KEY ORGANIZING QUESTION:
How do we and other animals learn?

KEY COMPETENCES	KEY CONCEPTS AND CONTENT	PERFORMANCE TASKS
Identify Predict Collect Analyze Write Explain	Systems Patterns of Change It is an internal process when new knowledge is constructed. We can, in part, understand what knowledge has been constructed by having the person or animal perform familiar tasks on which the new knowledge should be applied and observe any changes in the performance of those tasks.	**PERFORMANCE TASK I:** You and a teammate must identify a pet you can use to teach a new trick. Then think of an appropriate trick to teach that pet. Discuss different methods you might use to teach this trick. Select the method you will try and predict possible results. Practice teaching the pet to do the trick using your method and keep track of your results. How much time did you spend? How many times did you try the method? How did the pet react? Explain and exchange your method and your results with members of another team. Compare and contrast the two methods used for teaching the tricks. Write a note to the pet's owner explaining your method and the results. **PERFORMANCE TASK II:** Think of the way that you learn the spelling of new words. Discuss and compare your approach with the approaches of several of your friends. Select one method for learning new spelling words and predict how well you might do with this method. Try the method for 3 weeks and record your results. Then exchange your method and results with others. Compare and contrast the methods and results. Write a note to your teacher and explain your method. Be sure to include evidence of your results.

QUALITY CRITERIA:
"LOOK FORS"
- Clearly identify your method.
- Generate logical predictions.
- Collect and organize your data.
- Analyze the data by comparing and contrasting.
- Explain your findings in writing.

**Science:
Grade 5**

Performance Benchmark

LIFE SCIENCE: LIVING THINGS
CONTENT/CONCEPT STANDARD 1

KEY ORGANIZING QUESTION:
How do changes in environmental conditions affect living things?

KEY COMPETENCES	KEY CONCEPTS AND CONTENT	PERFORMANCE TASKS
Gather Compile Summarize Design Create Recommend Publish	Change Systems Stability Different organisms have different survival needs. Conditions in particular habitats can limit what kind of living thing can survive. Changes in an organism's habitat are sometimes beneficial to it and sometimes harmful.	**PERFORMANCE TASK I:** Many people take their pets along when traveling during hot weather. Gather information on how weather might affect pets in different conditions. Summarize the information you collect and identify the most important ideas for pet owners to know. Design and create a flyer on pet safety. Be sure to include the important ideas you have learned about the weather conditions and how they can affect a pet in different circumstances (for example, being in a parked car). Provide suggestions for the safety of pets traveling with humans. Publish your flyer and distribute it to local pet stores or owners. **PERFORMANCE TASK II:** Your family is planning a camping vacation to the Sonora Desert in the southwest United States. Your job is to gather information on the possible conditions you might encounter during your stay and determine the items you might need to take to ensure a safe camping vacation. Consider the time of year, the weather conditions, and the types of clothing (including footwear), shelter, and emergency supplies needed for this trip. Gather all the necessary information, summarize the important ideas, and then create a list of recommendations for a safe trip. Design and create flyer that explains your recommendations. Publish the flyer and distribute it through the local camping store.

**QUALITY CRITERIA:
"LOOK FORS"**
- Gather and compile appropriate information.
- Summarize the main ideas.
- Predict the possible needs in given circumstances.
- Create logical and useful recommendations.
- Design and create a useful quality product.
- Distribute your product to an appropriate audience.

Science: Grade 8 — Content/Concept Standards for Science — Performance Benchmark

LIFE SCIENCE: LIVING THINGS
CONTENT/CONCEPT STANDARD 3

KEY ORGANIZING QUESTION:
How have organisms adapted to their environments? What are the evidence and the causes for these modifications?

KEY COMPETENCES	KEY CONCEPTS AND CONTENT	PERFORMANCE TASKS
Select Analyze Draw Conclusions Depict Explain Justify	Systems Patterns of Change: As changes occur in an environment, organisms react to the change. Patterns of changes exist, and examining patterns will help us reach logical conclusions. There is a relationship between energy availability, stability, and adaptations of living things.	**PERFORMANCE TASK I:** Living things are extremely diverse. This diversity may be attributed to adaptations to their changing ecosystem and the ability of the species to survive dramatic changes. Select an environmental disaster that results in dramatic changes in the ecosystem (earthquake, flood, volcano, drought, fire). Describe how the changes in the environment affect plant and animal life. What survives and what doesn't? What adaptations do the plants and animals make to survive? Support your position with a scientific explanation. Create a visual depiction of the event and the adaptation. Prepare a presentation for your peers explaining your depiction and your position. **PERFORMANCE TASK II:** Think of an experiment that you could design to illustrate how plants are affected by changes in their environment. Describe the experiment, explaining what you would do and what the results might be. Then predict what would happen if your "experiment" occurred in your geographic region. What would happen? Which plants would survive and which would not? Create a depiction of your premises. Prepare a presentation for your peers explaining it and your position. Support your predictions with specific ideas and scientific principles.

QUALITY CRITERIA:
"LOOK FORS"

- Accurately describe an environmental disaster/experiment and the effect on plants and animals.
- Include accurate and specific scientific evidence to support conclusions.
- Organize the data clearly.
- Draw logical conclusions.
- Include accurate, logically sequenced details in visual depiction.
- Plan a presentation and deliver it to your peers.

Science: **Performance**
Grade 12 **Benchmark**

LIFE SCIENCE: LIVING THINGS
CONTENT/CONCEPT STANDARD 3

KEY ORGANIZING QUESTIONS:
How concerned should we be about microbes causing food spoiling or poisoning?

KEY COMPETENCES	KEY CONCEPTS AND CONTENT	PERFORMANCE TASKS
Investigate Observe Compare Question	Systems Organisms The positive and negative effects of the nature of organisms in the production, processing, handling, and serving of food	**PERFORMANCE TASK I:** You are to compare the effects of the microbes that cause food spoilage and food poisoning and determine how each condition can be prevented or minimized. Develop an inspection plan for the school cafeteria. Include in this plan what specific microbiological tests are to be performed, the questions that you would ask of the people who work there, and what specific things you would inspect in order to certify that the cafeteria is a safe place to prepare and serve food. Create a method for evaluating the results of the inspection plan of the cafeteria, specifying the criteria for judging it as being either safe or unsafe. Develop a written report on your process and findings and present it to the cafeteria personnel.

PERFORMANCE TASK II (continued):
serving site. Review your guide and its information with a representative from the food council. Make necessary changes; then publish and distribute through the home economics for family living courses.

QUALITY CRITERIA:
"LOOK FORS"
- Accurately observe and describe food and eating conditions that ensure human well-being.
- Ask logical causal questions about food poisoning and spoilage.
- Gather all necessary information.
- Summarize general ideas.
- Create specific recommendations.
- Design and develop a usable plan or guide.
- Conduct a trial test of the plan or guide.
- Respond as necessary.
- Create a quality product.
- Design distribution system and distribute.

PERFORMANCE TASK II:
We often hear about family outings that result in the attendees suffering from food poisoning. Gather information on food spoilage and food poisoning. Determine the various effects microbes have on food in different conditions and what the results might be for anyone consuming that food. Translate this information into specific recommendations for the serving and consumption of safe food. Design and develop a guide that would be useful to families. Be sure to include specific microbiological tests that could be performed, questions people need to ask themselves, and what they need to consider in determining a safe food preparation and

(continues in left column)

Content/Concept Standards for Science 21

LIFE SCIENCE – ECOSYSTEMS

Content/Concept Standards

The places that people live, the organisms that share this place, and the relationships among the people, organisms, and the environment are now being studied by ecologists. The questions that they strive to answer involve the theories and concepts that follow. Understanding these concepts will allow the students to take their proper place in the world.

What students should know how to do by the end of Grade 3

Students should observe and learn how an organism in an area may change over time, how organisms vary from area to area, and how the environment can affect organisms, as well as how organisms affect the environment. They should be able to

1. Explore, investigate, and classify the observable properties of living and nonliving things in their environments
2. Explore and explain the interrelationships between the living and nonliving things in their environments
3. Explore, investigate, and classify the changes in the appearance and habits of living and nonliving things from one region to another within their environments
4. Explore, investigate, and classify the variations in the appearance and habits of living and nonliving things over various intervals of time

What students should know how to do by the end of Grade 5

Students will begin to look for common characteristics as a basis for classification. The relationships between predator numbers and prey numbers, environmental conditions, and numbers of organisms in simple ecosystems should be explored. They should be able to

1. Demonstrate the use of the rules of classification to make predictions
2. Explore and investigate biotic relationships
3. Research, explore, and propose explanations of the nature of population dynamics in the sense of "lots of aphids leads later to lots of ladybirds"
4. Research, investigate, and explain simple food chains as they relate to the plants, animals, and nonliving things within a simple habitat
5. Research and analyze how patterns of changes in the environment have led to the extinction and endangerment of plants and animals

What students should know how to do by the end of Grade 8

Students should investigate how certain organisms are adapted to their habitats, how there are structural similarities and differences between living and extinct species, and how organisms can change in response to their environment, both structurally (natural selection) and behaviorally (migration, etc.). They should be able to

1. Relate the patterns of environmental changes to the observed migration patterns and movement of plants and animals
2. Investigate and describe the relationship of an organ to its organism, and an organism to its habitat or environment
3. Explain and graphically represent biomass in relation to food chains
4. Conduct research to support the conclusion that fossil and geological information provides evidence that some organisms living long ago are now extinct and some were similar to existing organisms
5. Investigate the evidence to explain how organisms adapt to their environments (natural selection)

What students should know how to do by the end of Grade 12

Students should build a coherent and comprehensive concept of an ecosystem and the factors (especially human) that can affect its population levels, equilibrium, diversity, and succession. Students should incorporate the factors of uncertainty and synergy into this concept. They should be able to

1. Investigate the subtleties of ecological balance and how the choices that humans have made and are making will influence this balance
2. Investigate and compare the relative effects of competition and the dynamic mechanism of succession
3. Research and describe the nature of multiple equilibria in the environment and the difficulty of predicting the outcomes of changes in one factor
4. Examine and analyze the rules for classification to include rules for class infusion and rejection
5. Examine and investigate evidence for how change occurs and how change in systems and organisms occurs over wildly different intervals of time

Content/Concept Standards for Science

Science: **Performance**
Grade 3 **Benchmark**

LIFE SCIENCE: ECOSYSTEMS
CONTENT/CONCEPT STANDARD 1

KEY ORGANIZING QUESTION:
What environment is best for plants?

KEY COMPETENCES	KEY CONCEPTS AND CONTENT	PERFORMANCE TASKS
Identify Compare Design Develop Present	Observable properties of plants Plants have special features that allow them to live in different environments.	**PERFORMANCE TASK I:** Your school plans to build two large planters, 4' x 6', for the front of the building. One planter will be in a shady location, and the other planter will be in direct sunlight all day long. Your job is to identify the plants and flowers that will do well in these two planters. Design and develop a sketch for each planter containing the locations of the recommended plants and flowers. Present your proposal to the school principal or beautification committee. **PERFORMANCE TASK II:** Your teacher will place at different stations around the room actual plants, pictures of plants, or descriptions of plants. You are to use the methods developed by your team to decide which plants belong together according to their environmental needs. Create and illustrate a plan to make a certain environmental area in your community more beautiful. Include in the plan a description of the plant groups you would use. Explain their particular needs and how you would meet them. Present your plan to members of a local garden club.
QUALITY CRITERIA: **"LOOK FORS"** • Observe and describe characteristics of plants. • Collect, organize, and analyze data. • Draw and apply reasonable conclusions. • Create a plan.		

Science:
Grade 5

Performance Benchmark

LIFE SCIENCE: ECOSYSTEMS
CONTENT/CONCEPT STANDARD 5

KEY ORGANIZING QUESTION:
Why are some species of plants and animals becoming extinct?

KEY COMPETENCES	KEY CONCEPTS AND CONTENT	PERFORMANCE TASKS
Identify Gather Analyze Conclude Recommend Create Review	Systems Patterns of Change: Predator, prey, and food relationships Animal behavior changes in complex ways in response to changing environmental conditions.	**PERFORMANCE TASK I:** You are to investigate the extinction of a species. You should relate your investigation to one or more of four major mechanisms: 1) destruction of habitats, 2) introduction of foreign species, 3) extermination of predators, and 4) hunting by people for food and fashion. Gather enough information on your topic to form a solid base of data. Analyze your information for causes and effects, and develop specific recommendations based on your findings. Create a poster for the Environmental Protection Agency (EPA) or your local environmental center. Your poster should contain recommendations for protecting existing species and the reasons why. Present and review your poster with a team of your peers and make necessary changes before displaying it or submitting it to the EPA.

PERFORMANCE TASK II (continued):
some solutions or alternatives to alleviate these concerns. Create a brochure for the animal protection agency that could be distributed to homes in the area.

QUALITY CRITERIA:
"LOOK FORS"
- Clearly state your problem or purpose.
- Identify useful resources.
- Analyze the information for mechanisms that affect the survival of the targeted species.
- Draw reasonable conclusions.
- Design a detailed representation or draft of your product.
- Review with an audience and adjust as needed.

PERFORMANCE TASK II:
The local animal protection agency in Oakland County, Michigan, which is north of the city of Detroit, has been receiving complaints from local residents about the presence of raccoons on their property. The raccoons get into garages, basements, and window wells. They are found on decks and in gardens. Why do you think this has become such a problem? Why are the raccoons so abundant within the metropolitan area? Could their presence affect other animals? Gather as much information as you can on this problem. Organize your data and develop
(continues in left column)

Content/Concept Standards for Science   25

Science: **Performance**
Grade 8 **Benchmark**

LIFE SCIENCE: ECOSYSTEMS
CONTENT/CONCEPT STANDARD 2

KEY ORGANIZING QUESTION:
What is the relationship of an organ to its organism and an organism to its environment?

KEY COMPETENCES	KEY CONCEPTS AND CONTENT	PERFORMANCE TASKS
Investigate Collect Compare Contract Summarize Synthesize Design Develop Review Present	Systems Patterns of Change Evolution: Similarities among organisms are found in internal anatomical features, which can be used to infer the degree of relatedness among organisms and their adaptability to their environment.	**PERFORMANCE TASK I:** A local agency is designing an area to be used as an environmental center by students and other community members. You have been asked to investigate the needs and adaptability of a water turtle, a box turtle, and an aquatic frog for this center because these might be interesting creatures for students to observe. Collect information that will give you enough data to determine the observable characteristics of each of these creatures and what they need to survive in the local environment. Compare and contrast their similarities and differences. Summarize and synthesize your data, and then design and develop a chart clearly displaying the unique features and characteristics of each of these animals, their needs, and the possibility of survival in such an environmental center. Present your chart to another team in your class and seek their recommendations. Make necessary changes before submitting your final project. **PERFORMANCE TASK II:** Investigate and collect information on the special adaptations you have as a human being that allow you to survive in various environments. Select another common animal, such as a bird, and identify its special adaptations that allow it to survive in various environments. Compare and contrast your findings on the bird and the human. Summarize and synthesize your findings; then design and develop a simple picture book for use with 7- and

PERFORMANCE TASK II (continued):
8-year-old students to help them understand the special adaptations of each to various environments. Present your rough draft to a team of your peers and ask them to make recommendations. Edit as necessary and publish your book. Share it with a class of 7- or 8-year olds.

QUALITY CRITERIA:
"LOOK FORS"
- Identify a clear purpose.
- Collect accurate information from a variety of resources.
- Examine the main points of each topic.
- Select major similarities and major differences.
- Condense the information for use in your product.
- Represent your ideas using appropriate materials.
- Review and adjust according to need.

(continues in left column)

Science: **Performance**
Grade 12 **Benchmark**

LIFE SCIENCE: ECOSYSTEMS
CONTENT/CONCEPT STANDARD 1

KEY ORGANIZING QUESTION:

What are some of the possible choices that humans can make to maintain the ecological balance of an ecosystem?

KEY COMPETENCES	KEY CONCEPTS AND CONTENT	PERFORMANCE TASKS
Research Interpret Draw Conclusions Predict Results Explain	Energy Systems Patterns of Change Scale and Structure: Disturbing the ecosystem changes the energy flow. Demonstrates the interrelationship between the elements of the system Effects of a steady trend of disturbance to a system Local disturbances create global affects.	**PERFORMANCE TASK I:** The Potomac River is a complex, dynamic ecosystem that has experienced change. Consider the water (volume, quality, path), fish and plant life, and the area immediately surrounding the river, including plants and animals for whom this area is their native environment. Consider current use of the river by humans and the impact people have on this ecosystem. Given current conditions and human practices, predict what the river conditions will be in the next 5-10 years. What global impact will changes in the Potomac have? Develop a position about what actions the people of Washington, D.C., should be engaged in to maintain the health of the river and the plants and animals that live in this ecosystem. Develop charts and graphs to support your position. Share your findings and recommendations with the Potomac River Agency or a local environmental group.

QUALITY CRITERIA:
"LOOK FORS"

- Establish a clear purpose.
- Utilize a variety of resources.
- Accurately interpret data and information.
- Develop logical conclusions based on significant, accurately interpreted scientific principles.
- Support conclusion with specific, accurate data about both local and global environmental impact.
- Predict logically and show clear understanding of human impact on the environment.
- Organize ideas clearly and accurately for presentation (charts, graphs, etc.).

PERFORMANCE TASK II:

Personal and family decisions can have a profound environmental impact. Design and develop a "green plan" for a family of four that addresses what they buy, what services they use, and how they handle waste material. Predict the short- and long- term effect of these practices on the environment. Support your predictions with specific examples and relevant scientific data. Share your findings with a local environmental research group.

PHYSICAL SCIENCE – MATTER

Content/Concept Standards

Atomic theory describes and explains many natural phenomena. Students must investigate and learn the many properties of matter to be able to explain, understand, and predict how matter combines and responds to changes such as temperature, pressure, and so forth.

What students should know how to do by the end of Grade 3

Students should explore and investigate the observable properties of a wide variety of materials and learn how to categorize matter according to various properties. Students should explore how matter responds to simple changes involving mixing, bending, heating, freezing, scratching, and dissolving and be required to describe what was done and how the material responded to these changes. They should be able to

1. Describe and identify the observable properties of matter and demonstrate that the descriptions can be organized in different ways
2. Gather information about matter using various tools and instruments
3. Explore and describe various measures and concepts such as length and weight (mass)

What students should know how to do by the end of Grade 5

The study of matter should become more systematic and quantitative. Students should design simple experiments to explore, measure, collect data, and present their findings about the more complex properties such as buoyancy, density, heat flow, and so on. They should be able to

1. Research, create, and explain concrete models that describe simple atomic structure
2. Explore and investigate changes that occur in a variety of ways in matter and demonstrate that these changes can be categorized by the differences in chemical and physical properties
3. Explore and investigate simple aspects of the laws of conservation (mass, volume, number, etc.) of matter
4. Explore and investigate how new materials may be made by chemical reactions that combine two or more different elements (the new material can have properties different from the original materials)

What students should know how to do by the end of Grade 8

By this time students should have a sufficient grasp of the general properties and models of matter and understand that many phenomena can be explained—from solubility to chemical and physical changes—using the basic assumptions of the kinetic-molecular theory. Interrelationships among temperature, chemical changes, and physical changes should be investigated. Historical developments that have led to our present understanding of matter should be integrated. The concepts and properties of elements, molecules, mixtures, compounds, and the conservation of matter should be developed. Students should be able to

1. Examine the kinetic molecular theory to explain phenomena in simple correspondence with the theory, such as expansion due to greater vibration of particles and solutions as the intermingling of particles
2. Examine, analyze, and integrate information on the structure of the atom and the properties of the particles of the atom with daily living
3. Compare and contrast the properties of solids, liquids, and gases in terms of observable characteristics and those that are not directly visible
4. Explore and investigate the concept of force, and distinguish among the various types, such as adhesion, cohesion, capillary, surface tension, gravity, and friction (i.e., can distinguish between force and pressure)

5. Investigate the behavior of gases and describe qualitatively the concepts of compression, expansion, and temperature
6. Examine the concept of temperature as it relates to matter, human perceptions, and the kinetic molecular theory
7. Investigate simple chemical reactions and classify the type of reaction and the nature of the product

What students should know how to do by the end of Grade 12

Students will continue to investigate the complexities of matter to the atomic and nuclear levels. Fully developed should be the concepts of the structure of matter and the forces that mediate that structure, from chemical and nuclear bonds to gravitational and electromagnetic forces. Applications of these concepts to the world around the student and the use of technology should be integrated. They should be able to

1. Explore, investigate, and classify the various stable arrangements of matter including atoms, compounds, molecules, and mixtures
2. Investigate the observable properties of substances bound by various types of chemical bonds
3. Research and investigate how the configuration of atoms and arrangement of electrons within the atoms determines the various physical and chemical properties of matter
4. Apply the concepts of conservation of matter to explain a variety of phenomena, from chemical and nuclear reactions to the swinging of a pendulum
5. Distinguish between organic and inorganic molecules and compounds in terms of structure, function, and application

Science: Grade 3

Performance Benchmark

PHYSICAL SCIENCE: MATTER
CONTENT/CONCEPT STANDARD 2

KEY ORGANIZING QUESTION:
How and why are common measuring instruments used?

KEY COMPETENCES	KEY CONCEPTS AND CONTENT	PERFORMANCE TASKS
Investigate Observe Classify Record Compare Explain	Scale and Structure Patterns of Change: The usefulness of mathematics in science and technology will be clear to students if they experience it often. Comparison is made through weight (heavier/lighter), dimensions (longer/shorter), and temperature.	**PERFORMANCE TASK I:** (Notes to the teacher: ask students to bring in a variety of objects. These objects should range in length or height from 1 or 2 inches to approximately 18 inches. These objects might include familiar household items such as a Coke® can, a key, a cup, a bag of M&M's®, and so on. The teacher should also contribute a variety of items to this collection if necessary. Also needed for this performance are a number of containers of water with various temperatures placed at different locations in the room. Last, students should have access to a variety of measuring instruments.)

PERFORMANCE TASK II (continued):
how you use them to collect information on various objects. Measure the selected objects you have identified in your home and record the length, width, height, and weight. Measure different liquids found in your home and record the temperatures. Be sure to include a description of the location. Analyze your chart. What connections can you make between the different properties? Take your information to school. Meet with a partner and share your data and your findings. Explain how the measuring tools helped you gather information at home.

QUALITY CRITERIA:
"LOOK FORS"
- Observe and accurately record.
- Organize information into categories.
- Select similarities.
- Explain connections to a partner.
- Explain how the measuring tools helped you gather information.
- Explain other ways to use measuring instruments.

You are going to use a variety of measuring instruments to examine the properties of weight, dimensions, and temperature of various objects. Measure the selected objects and record the length, width, height, and weight. Visit each water container and then read and record the temperature. Also record a description of the location (in sunny window, in a closet on floor, or in an ice chest). Record your gathered information on a chart and analyze the results. Can you make any connections between the different properties? Explain to a partner how the different instruments helped you gather the information. How could you use these instruments in a different setting?

PERFORMANCE TASK II:
Take a set of these measuring instruments home and demonstrate to an adult there
(continues in left column)

Science: Grade 5

Performance Benchmark

PHYSICAL SCIENCE: MATTER
CONTENT/CONCEPT STANDARD 2

KEY ORGANIZING QUESTION:
How are chemical and physical changes differentiated?

KEY COMPETENCES	KEY CONCEPTS AND CONTENT	PERFORMANCE TASKS
Observe Investigate Analyze Classify Conduct Explain Conclude	Systems Patterns of Change Evolution: Things change in steady, repetitive, or irregular ways—or sometimes in more than one way at the same time.	**PERFORMANCE TASK I:** You and your team are to watch your teacher demonstrate some changes from the cleavage of calcite (or other crystalline substance with cleavage planes), the burning of paper, a match, a candle, and the mixing of clean sand with water. Discuss these changes and develop a definition for the two different types of changes (chemical and physical) that you have observed. Next you are given some weak vinegar, baking soda, salt, sugar cubes, and water. Decide on a method to mix these items by pairs; record and describe what happens. Decide which of the above definitions fits each pairing of substances and report your conclusions to the class.

QUALITY CRITERIA:
"LOOK FORS"
- Accurately observe and describe.
- Ask logical causal questions.
- Recognize, generate, and state alternative explanations.
- Generate logical predictions.
- Conduct the experiment to test your prediction.
- Draw and apply reasonable conclusions.

PERFORMANCE TASK II:
You will be given salt, water, ice chips, and a thermometer. Measure the temperature of each separately and describe each. Mix them together by pairs. Measure the temperature and describe any changes. Use the definition created by your group in the above performance task to classify each change. Return to your group and compare your notes. Come to an agreement on the classification and write an explanation of your conclusions. Read it out loud to the rest of the class.

Science: Grade 8

Performance Benchmark

PHYSICAL SCIENCE: MATTER
CONTENT/CONCEPT STANDARD 1

KEY ORGANIZING QUESTION:
How can we use models to explain correctly how physical changes occur from an atomic perspective?

KEY COMPETENCES	KEY CONCEPTS AND CONTENT	PERFORMANCE TASKS
Examine Select Review Create Organize Explain	Systems Patterns of Change Evolution: Physical and biological systems tend to change until they become stable and then remain that way unless their surroundings change. A system may stay the same because nothing is happening or because things are happening but exactly counterbalance one another.	**PERFORMANCE TASK I:** Your group is given a variety of construction materials, from toothpicks to cotton balls. Examine all the materials and select those materials that will be most useful in creating a model or a series of models. The physical model or sequence of models should represent the various physical changes from the cleaving of a crystal, the expansion of a gas, to the dissolving of sugar in water. Create your model, and then present it to another group and explain how it represents a physical change. Be sure to include some features about the model that represent changes other than physical.

QUALITY CRITERIA:
"LOOK FORS"
- Identify your task and purpose.
- Select appropriate materials.
- Review possible alternatives for your model.
- Decide on best physical representation.
- Build your idea using your selected materials.
- Organize your ideas for presenting.
- Explain and justify your model and process.

PERFORMANCE TASK II:
Your group must meet and discuss the physical changes that can occur in the kitchen when someone is preparing a meal. Examine all the possibilities and select the best examples of physical change that occurs. Create a model or a series of models to represent your selections. Present your model to another group and explain how it represents a physical change. Take your model home and explain it and the connections to preparing a meal to someone there. Ask the person to respond to your ideas.

Science: **Performance**
Grade 12 **Benchmark**

PHYSICAL SCIENCE: MATTER
CONTENT/CONCEPT STANDARD 2

KEY ORGANIZING QUESTION:
How can events in our daily lives be related to particles of the atom?

KEY COMPETENCES	KEY CONCEPTS AND CONTENT	PERFORMANCE TASKS
Select Examine Share Compare Create Relate Present Exchange	Systems Patterns of Change Evolution: Atoms are made up of a positive nucleus surrounded by negative electrons. An atom's electron configuration, particularly the outermost electrons, determines how the atom can interact with other atoms. The rate of reactions among atoms and molecules depends on how often they encounter one another, which is affected by the concentration, pressure, and temperature of reacting materials.	**PERFORMANCE TASK I:** Determine the type of bond that is responsible for water sticking to the tip of your finger or for making your hair go limp when it is wet. Explain and compare your ideas with a teammate and then together create a model to represent atomic configuration. Present and exchange your model and a clear explanation with another team. What conclusion can you reach as a group? **PERFORMANCE TASK II:** You are given a variety of safe compounds and some tools such as a conductivity meter, hardness test plate, and so forth. You are to choose two or more of the compounds and investigate the properties of each. Explain and discuss your findings with your teammate and together create a method for grouping the materials by common properties. Then create a model to represent the interaction of these properties. Relate this process to a common daily example. Present and exchange your ideas, your model, and your common examples with another team. What conclusions can you reach as a group?

QUALITY CRITERIA:
"LOOK FORS"
- Identify a clear purpose and procedure.
- Survey available materials or ideas.
- Organize according to purpose.
- Include specific and accurate details.
- Elicit feedback and respond.
- Select commonalities.
- Review possible alternatives for a physical representation.
- Select a plan.
- Connect theory to reality.
- Evaluate audiences' response and adjust accordingly.
- Share conclusions.

PHYSICAL SCIENCE – ENERGY

Content/Concept Standards

Energy is a difficult concept for even the scientist to define. Analogy is often helpful, and descriptions of the many events that require energy or use energy are useful if students are to construct a personal definition and concept of energy. Energy has many forms and comes from many sources, so it is important for the teacher and the students to focus on the observable qualities related solely to energy. Of the six or so conservation laws, the conservation of energy is one that is at the heart of thermodynamics and has many practical applications. Other energy-related ideas that students must understand are that all physical events require the transfer of energy from one form to another and from one object to another, and not all energy is useful energy.

What students should know how to do by the end of Grade 3

At this age, students benefit from talking about energy before any attempt is made to define energy. To realize that we get energy from the sun and from our food and that moving things require energy is sufficient. They should be able to

1. Recognize a variety of ways of making things move
2. Recognize that overuse or misuse can lead to depletion of resources
3. Recognize the sun as a source of heat for the earth

What students should know how to do by the end of Grade 5

Heat is introduced here as the sole, introductory form of energy because students are familiar with heat and its many sources. They should be able to

1. Explore and classify the many sources of heat
2. Explore and investigate how heat can flow from place to place and from object to object, and that this heat flow can be indirectly measured by the change of temperature of the objects
3. Demonstrate how various materials can serve as conductors and how might they affect the flow of heat
4. Explore various materials as insulators
5. Explore and describe the condition of thermal equilibrium (a dynamic process)
6. Investigate the effects of heat on various substances in terms of reaction to and the need for heat (burning of paper and the freezing of water) in the requirements of living systems

What students should know how to do by the end of Grade 8

Forms and sources of energy other than heat are appropriate to be studied. Energy transfer from place to place and object to object can help introduce the concept of conservation. Energy sources must be distinguished from energy forms. Students should be able to

1. Investigate and classify various phenomena that illustrate the concepts of kinetic and potential energy
2. Investigate and explain that energy can be converted from one form to another
3. Investigate and observe the ways energy is transformed and why it needs to be transformed
4. Investigate and explain the importance of energy conservation, research the ways energy is being conserved and wasted, and propose a reasonable solution to reduce the waste
5. Observe and explain the obvious and subtle changes in the physical and chemical properties of a substance as energy is absorbed by or removed from matter (expansion, burning, phase change)
6. Analyze the concept of the mechanical equivalent of heat in terms of its scientific and practical importance

What students should know how to do by the end of Grade 12

The study of heat becomes more complex and quantitative. Mechanical equivalent of heat, and calculations of elastic, gravitational potential, kinetic, and rotational energies are calculations that will allow students to test the concepts that they have constructed so far. Measurements of temperature and the calculations of heat flow and latent heat along with those of electrical energy and power provide a convenient way to relate electrical to thermal energy. Atomic and molecular structures and their relationship to radiant energy should also be investigated. Nuclear energy should also be introduced. Students should be able to

1. Distinguish between heat and temperature
2. Relate temperature to the response by matter to the flow of energy
3. Investigate and classify the factors that influence the rate of chemical reactions
4. Measure and describe how electrical energy flows through AC and DC circuits.
5. Observe and assess the properties and behavior of waves in various substances
6. Design and construct a device that applies the concepts of simple machines and conversion of energy to do a measured amount of work to accomplish a useful task
7. Apply the concepts of mechanics and dynamics to analyze and describe the causes of the motion of a real object

Content/Concept Standards for Science

Science:　Grade 3　　**Performance Benchmark**

PHYSICAL SCIENCE: ENERGY
CONTENT/CONCEPT STANDARD 1

KEY ORGANIZING QUESTION:
What makes things move?

KEY COMPETENCES	KEY CONCEPTS AND CONTENT	PERFORMANCE TASKS
Predict Observe Describe Categorize Explain	Systems Cause and Effect: Energy is the ability to do work or make things move.	**PERFORMANCE TASK I:** As a scientific investigator, you are going to join your class on a short walk around the school. Before you go, you must predict the kinds of things you will observe that move. On your walk observe and record the things you see that move and determine what causes them to move. When you return to your room, cluster or categorize the items you observed according to what caused their movement. Draw a picture to show your categories. Explain what makes objects in your different categories move. Describe the energy transformation or energy conversion that takes place. Collect your individual sheets of pictures and bind them into a big book to be enjoyed by the class or read to a group of second graders. **PERFORMANCE TASK II:** The teacher will give you 12 pictures that depict things from home, school, or nature. All of the things in these pictures move. Before looking at the pictures, predict what causes each item to move. Group the 12 pictures into categories according to the cause of their movement. Explain to a learning partner what causes them to move, and then draw a picture of some other things you know that fit into each of the categories you identified. Write a sentence about each one telling what causes it to move, and describe the energy transformation or conversion that takes place.
QUALITY CRITERIA: **"LOOK FORS"** • Generate logical predictions. • Observe to gather information. • Analyze and organize the data. • Create a depiction with categories. • Explain your impressions and ideas.		

Science: Grade 5

Performance Benchmark

PHYSICAL SCIENCE: ENERGY
CONTENT/CONCEPT STANDARDS 3 & 4

KEY ORGANIZING QUESTION:
How effective are various materials as insulators or conductors?

KEY COMPETENCES	KEY CONCEPTS AND CONTENT	PERFORMANCE TASKS
Experiment Observe Draw Predict Discuss Recommend Publish	Systems Energy Stability: The nature of conservation Various materials have different insulating properties. Various materials serve differently as conductors of heat.	**PERFORMANCE TASK I:** As a team, you are to conduct a simple test to see how well different materials conduct heat. You will need a glass beaker, beads, butter, a pan of hot water, a wooden spoon, and a paper drinking straw. Stick a small bead on each test item using a small blob of butter. Make sure the bead is the same distance from the far end on each item. Make all the blobs of butter the same size. Stand the items upright in the beaker and pour in about 3 inches of hot water. Heat from the water does what? What happens to the blob of butter and the bead? What have you learned from this experiment? Based on your findings, draw an outline of a house including roof, doors, and windows. Predict the percentage of heat that will escape from the house and where it will escape. Discuss how this heat loss problem could be solved. Create a list of possible solutions for the losses and share them with another team. Publish the possible solutions and share them with your parents. Discuss whether any of your solutions could apply to your home.

PERFORMANCE TASK II (continued):
problem could be solved. Create a list of possible solutions for the losses and share them with another team. Publish a newsletter containing your recommendations for the school, and share it with others in the school community. Discuss the reactions to your newsletter.

QUALITY CRITERIA:
"LOOK FORS"
- Experiment with various materials as heat conductors.
- Observe the results of the experiment.
- Transfer your findings to predictions on heat loss from a building.
- Discuss various solutions.
- Select the best ones.
- Publish your recommendations for the problem.
- Share reactions.

PERFORMANCE TASK II:
Working with different partners, conduct the above experiment again and compare your results with the first test. Based on your findings, create an outline drawing of your school including roof, doors, and windows. Predict the percentage of heat that will escape from the school and where it will escape. Discuss how this heat loss

(continues in left column)

Content/Concept Standards for Science

Science: Grade 8

Performance Benchmark

PHYSICAL SCIENCE: ENERGY
CONTENT/CONCEPT STANDARD 5

KEY ORGANIZING QUESTION:

How do physical properties change when heat is added to or removed from a system?

KEY COMPETENCES	KEY CONCEPTS AND CONTENT	PERFORMANCE TASKS
Investigate Analyze Classify Identify Justify Explain Share	Energy Systems Patterns of Change Stability: Atoms and molecules are perpetually in motion. Increased temperature means greater average energy of motion, so most substances expand when heated. In solids, the atoms are closely locked in position and can only vibrate. In liquids, the atoms or molecules have higher energy of motion, are more loosely connected, and can slide past one another; some molecules may get enough energy to evaporate into a gas. In gases, the atoms or molecules have still more energy of motion and are free of one another except during occasional collisions.	**PERFORMANCE TASK I:** Your team has been given the task to investigate how the addition or removal of heat affects a system. Fill a shallow pan part way with water and place a water glass upside down in the pan. Discuss how the trapped air in the glass can be heated or cooled. Heat and cool the air inside the glass and observe the changes that occur. Classify the change as either a physical or chemical change and give reasons to support your answer. Discuss and describe where in nature this kind of change occurs naturally. How would this change relate to inflating bike tires? Write a note explaining your findings to fifth graders. Explain why this is important for them to know. **PERFORMANCE TASK II:** When bread is made at home or at the bakery, bubbles of gas are responsible for one of the properties of the bread. Explain this property and describe how it happens. Explain to your parents why the freezer door is hard to open immediately after it has been closed. How does a living plant respond to a change in heat from its environment? Describe other physical properties that change as heat is added or removed from a system. Explain in a letter to your teacher the relationship among these items.

QUALITY CRITERIA:
"LOOK FORS"

- Identify your purpose.
- Carefully conduct an experiment.
- Collaborate on the findings.
- Identify the type of change you observed.
- Justify your position.
- Relate your findings to one that occurs in nature.
- Explain your understandings and connections in writing.
- Share with a specific audience.

Science: **Performance**
Grade 12 **Benchmark**

PHYSICAL SCIENCE: ENERGY
CONTENT/CONCEPT STANDARD 6

KEY ORGANIZING QUESTION:
How does force affect motion?

KEY COMPETENCES	KEY CONCEPTS AND CONTENT	PERFORMANCE TASKS
Recall Gather Discuss Analyze Justify Support Review	Energy Patterns of Change Stability: Energy is necessary to cause a change. Newton's Laws are invariant to Lorentz and time invariant. A static condition results if the sum of all forces and torques are zero.	**PERFORMANCE TASK I:** Discuss in terms of Newton's laws of motion: a. Why people in a moving car lurch forward when the car suddenly slows down b. What happens to the passengers of a car that makes a sharp, quick turn Apply this information to the safety design of a vehicle. Present and explain your design to another team. Respond to their input and share your final copy with an authentic design team from one of the automotive companies. Seek their reactions.

QUALITY CRITERIA:
"LOOK FORS"

- Identify your purpose.
- Review the theory and problem.
- Select the most appropriate information for the purpose.
- Prioritize the information.
- Create a design and written explanation.
- Include necessary supporting details.
- Review and edit as necessary.

PERFORMANCE TASK II:
A friend tells you that when riding in a bus, she was badly bruised by slamming against the side of the bus during a sharp turn. She is quite upset and angry with the bus driver. In terms of Newton's laws of motion, develop an explanation for your friend so she can understand why this happened. Put your explanation in writing and support it with specific examples. Be sure to include suggestions that might serve to prevent a repeat occurrence.

EARTH SCIENCE – METEOROLOGY

Content/Concept Standards

One of the underlying scientific themes that shows the relationship between the living and nonliving worlds is the study of meteorology and its effects on living things. This unit provides for observations of patterns and relationships between meteorological conditions and processes, as well as the linkage between weather, living things, and the behavior of humans.

What students should know how to do by the end of Grade 3

Students should observe and record weather changes, recognize patterns in these changes, and relate these changes to living and nonliving things in their environment. They should be able to

1. Record and compile observable weather conditions, and search for simple relationships and patterns in the weather over time intervals of hours, days, and months
2. Investigate the simple behaviors exhibited by living things in response to changes in the weather
3. Explore and describe the observable effects of weather and changes in the weather on nonliving things

What students should know how to do by the end of Grade 5

Students should conduct studies of various long-term weather conditions, relate seasonal cycles to local weather, and explore how changing conditions affect processes such as evaporation and condensation. They should be able to

1. Measure, record, and summarize the results of short- and long-term weather studies
2. Investigate and relate the various cycles of weather to local environmental conditions
3. Investigate the factors that affect processes such as evaporation and condensation

What students should know how to do by the end of Grade 8

Students will use various instruments and technologies to collect data and make predictions. Students will construct several basic weather measurement instruments and conduct investigations of the relationships between changes in weather and changes in living systems. They should be able to

1. Use various instruments and data sources to predict short- and long-term weather conditions
2. Investigate the multitude of observational systems used in forecasting weather
3. Investigate relationships between weather and the behavior of living systems

What students should know how to do by the end of Grade 12

Students will research and investigate the relationships between the physical world (climate, geography, etc.), the occurrence of natural disasters, and the locations of human population centers. In addition, students will explore and investigate local bodies of water to show how their characteristics allow current life-forms to exist. They should be able to

1. Analyze relationships between the physical world, natural disasters, and their effects on humans
2. Investigate the importance of water vapor in the atmosphere and its relationship to weather
3. Investigate the troposphere and the use of satellite images in weather forecasting

Science: Grade 3
Performance Benchmark

EARTH SCIENCE: METEOROLOGY
CONTENT/CONCEPT STANDARD 2

KEY ORGANIZING QUESTION:
How does temperature affect the simple responses of living things?

KEY COMPETENCES	KEY CONCEPTS AND CONTENT	PERFORMANCE TASKS
Observe Examine Collect Discuss Describe Record Recommend Organize Publish	Patterns of Change: Things can change in different ways, such as size, weight, color, and movement. Some small changes can be detected by taking measurements.	**PERFORMANCE TASK I:** You are to investigate the effects of temperature on the breathing rate of fish. You will go to four stations. Each station will have a thermometer and a goldfish in a glass container of water. At each station, record the temperature of the water and observe the behavior of the goldfish. Develop a chart or graph and record the results of your observations. Draw conclusions about the relationships you have observed. Develop an illustrated and written guide for a pet store to give to people who purchase goldfish. Include the results from your investigation and what this means to fish owners. **PERFORMANCE TASK II:** You are to investigate the effects of temperature on human bodies engaged in running. You and two teammates are to collect data from your own experiences. Select a place in the sun to run for 4 minutes. Before running, take the temperature of each person and record your findings. Also record the pulse rate of each person. When this is completed, set a timer and begin your 4-minute run in the sun. As soon as you stop, take the temperature and check the pulse rate of each person. Record this information. Rest for 10 minutes and then find a cool shady place (or a cool indoor area) where you might run again for 4 minutes. Before running, take and record the temperature of each person and also take and record their pulse rate. Now run in a cool place for 4 *(continues in left column)*

PERFORMANCE TASK II (continued):
minutes. Immediately take your temperatures and pulse rates. Record this new batch of data. Analyze your collected data and determine if there was any effect of temperature on the human body. Draw conclusions on the relationships you discover. Develop an illustrated and written guide for joggers that contains information and helpful hints.

QUALITY CRITERIA:
"LOOK FORS"
- Clearly state your purpose.
- Accurately observe and describe the effect of temperature on breathing patterns.
- Collect, organize, and analyze data.
- Draw reasonable conclusions.
- Organize information logically in a guide.
- Support with details.
- Publish the guide and distribute.

Content/Concept Standards for Science

Science: **Performance**
Grade 5 **Benchmark**

EARTH SCIENCE: METEOROLOGY
CONTENT/CONCEPT STANDARD 3

KEY ORGANIZING QUESTION:
What are the conditions that affect evaporation?

KEY COMPETENCES	KEY CONCEPTS AND CONTENT	PERFORMANCE TASKS
Observe Describe Explain Predict Collect Organize Conclude Procedure	Energy Systems Scale: Energy is necessary to make matter move. Evaporation is a part of the water cycle. There is a relationship between what is learned in the lab and occurrences in the world.	**PERFORMANCE TASK I:** You often notice water evaporating (e.g., puddles of water disappearing after a rain), and you know that on some days water evaporates faster than others. Your team is to investigate the factors that affect the speed of evaporation using the following equipment: **Thermometer** **Lamp** (60W bulb) as a safe source of heat **Cardboard** for fanning the damp sponge **Sponge** - household type (LWH ≈ 1cm x 7cm x 14cm) cut into 4 equal pieces **Balance** (0.1 gm accuracy) **Small container** for adding water to the sponge ℵ 5ml Investigate the factors of temperature and wind, one at a time, and discuss your results. If you were a teacher getting ready *(continues in left column)*

PERFORMANCE TASK I (continued):
to take a field trip to a neighborhood park, how might these factors affect the trip? Write a note to parents and give advice on preparing their children for a hot, windy day.

QUALITY CRITERIA:
"LOOK FORS"
- Accurately observe and describe factors that affect evaporation.
- Recognize, generate, and state alternative explanations for evaporation.
- Generate logical predictions related to specific situations.
- Collect, organize, and analyze data.
- Draw and apply reasonable conclusions.

PERFORMANCE TASK II:
Conduct the above investigation on the factors that affect the speed of evaporation. Discuss your results and relate them to natural weather conditions. Predict how these factors might affect a farmer, his farmland, and his crops. Discuss and describe the possibilities. Gather and organize your information. Write an article for a local farm journal on your findings. What advice could you offer the farmers.

Science: Grade 8

Performance Benchmark

EARTH SCIENCE: METEOROLOGY
CONTENT/CONCEPT STANDARD 2

KEY ORGANIZING QUESTION:
How can various tools help predict changes in weather?

KEY COMPETENCES	KEY CONCEPTS AND CONTENT	PERFORMANCE TASKS
Construct Observe Record Compare Interpret Conclude Summarize Create Present Explain	Systems Patterns of Change: Air pressure changes with atmospheric conditions. Changes can give a good indication of weather to come.	**PERFORMANCE TASK I:** You will need: glass, balloon, rubber bands, adhesive tape, straw, wooden base, card. Stretch the balloon over the glass, and secure it tightly with rubber bands. Attach the straw to the top of the balloon, and cut the opposite end into a point. Fix the glass to the wooden base with adhesive tape. Fold the card at the bottom to make a stand, and fix it to the wooden base to align with the end of the straw. Mark the level of the straw on the card each day to record the changes in air pressure. Observe the changes on your barometer. Carefully record your observations. Be sure to compare information from your barometer to information from a professional barometer. What can you learn from these observations? What information have you recorded? Is this useful information? Who would be interested in it? Why? At the end of a week, summarize your data and create a presentation for your classmates. Explain your findings and your conclusions.

PERFORMANCE TASK II (continued):
from this information? How would they benefit? After a week, summarize your data and create a presentation for your classmates. Explain your findings and your conclusions.

QUALITY CRITERIA:
"LOOK FORS"
- Accurately observe and describe the reactions of living things to changes in the weather.
- Ask logical causal questions on the relationships between behavior and weather.
- Recognize, generate, and state alternative explanations.
- Generate logical solutions for ensuring the survival of the largest number of animals.
- Organize and analyze data relevant to behavior and seasonal changes.
- Draw and apply reasonable conclusions.

PERFORMANCE TASK II:
Conduct the above experiment again, only this time be sure to check and record the temperature each time you check and record the information from your barometer. What other information could you collect that would be helpful in predicting changes in weather? Do it. Carefully record your observations. What are you learning from these observations? Is this useful information? Who could benefit
(continues in next column)

Content/Concept Standards for Science

Science: Grade 12

Performance Benchmark

EARTH SCIENCE: METEOROLOGY
CONTENT/CONCEPT STANDARD 1

KEY ORGANIZING QUESTION:
How are people affected by changes in weather systems?

KEY COMPETENCES	KEY CONCEPTS AND CONTENT	PERFORMANCE TASKS
Research Analyze Discuss Categorize Write	Energy Systems Patterns of Change Scale: Many natural events are the result of significant energy being unleashed. Changes in one element of a system influence other elements of the system. Many natural hazards are cyclical events. Not all elements of a system are equally affected.	**PERFORMANCE TASK I:** You are a member of a city advisory flood control team. Your task is to research and analyze the political, social, and individual effects of flooding on your community or a particular community of your choice. Read the article entitled "The Broken Heartland" (*Time,* August 9th, 1993) on the 1993 flooding of the Mississippi basin. Discuss and categorize the effects of the flooding. Write a memorandum to the city council recommending possible solutions for preventing or minimizing the harmful effects of such a flood in your area. **PERFORMANCE TASK II:** Your family lives in a flood basin. Research and analyze the possibilities for flooding that could occur. Discuss and categorize the effects of such a flood with a teammate. Write a detailed plan with activities and dates that describes the decisions that you would make for the well-being of your family in preparation for a flood. Provide support for your decisions.

QUALITY CRITERIA:
"LOOK FORS"

- Clearly state your purpose.
- Utilize a variety of resources.
- Prioritize the gathered information.
- Select the most appropriate for your purpose.
- Exchange ideas with a teammate.
- Reflect, review, and organize according to essential categories.
- Create a written document for your audience.

EARTH SCIENCE – GEOLOGY

Content/Concept Standards

The area in which you live is the result of many different processes and forces that have shaped the land. This substrand explores these processes and forces and develops the ideas of classification, observation, comparison, and adaptation. In addition, this substrand shows how humans have affected their environment and how the environment has affected them.

What students should know how to do by the end of Grade 3

Students should observe and explore their area and compare it to other areas. Various local rocks and minerals should be collected to show simple similarities and differences. Students should become familiar with which things change and what seems to cause change. They should be able to

1. Explore the local environment and describe the features that make it similar and dissimilar to deserts, forests, and other simple land and water forms
2. Examine and investigate various rocks and minerals from the local environment and determine, in simple ways, how they are alike and different
3. Recognize that chunks of rock come in many sizes and shapes, from boulders to grains of sand and even smaller
4. Observe that animals and plants sometimes cause changes in their surroundings

What students should know how to do by the end of Grade 5

In these years, students should acquire more information about the physical environment, becoming familiar with the details of geological features. A trench or crosscut should be observed, characterized, and classified by the students. Investigations of landforms and layering should be performed to see the evidence for geological change over time. Students should be able to

1. Examine a trench in the soil and describe the characteristics that differentiate the various layers
2. Research and investigate the clues that support the conclusion that rocks, minerals, soils, and land features have changed over time
3. Collect and examine various soil samples and develop a system to classify soils
4. Observe and map locations of hills, valleys, rivers, and other land features
5. Note variety among various geological components
6. Observe the elementary process of the rock cycle—erosion, transport, and deposit
7. Investigate different combinations of minerals as well as soil

What students should know how to do by the end of Grade 8

Students should observe characteristics of rocks of different types and classify how natural forces and energy sources have affected living and nonliving things. Human adaptation to various natural forces should be studied. They should be able to

1. Research and analyze the various natural forces and sources of energy and classify their effects on the living and nonliving things
2. Explore and examine how humans can adapt to the forces of nature
3. Examine the properties of various minerals and classify them as being sedimentary, metamorphic, or igneous and give evidence concerning the historical geology of their local environment
4. Research and discuss the various clues to identifying the characteristics of the earth's structure

What students should know how to do by the end of Grade 12

Students should explore the relationship among isostasy, equilibrium, uniform processes, and the geologic features that result. They should develop a hypothesis concerning the internal structure of the earth and defend it with credible scientific data. The relationship between human activity and depletion of resources should be investigated. Students should be able to

1. Investigate the evidence that relates humans' activities to the resources of their environment
2. Develop reasonable explanations for the earth being composed of solids, liquids, and gases and use credible evidence to support the explanations
3. Observe, collect data, and construct models to show that the earth's interior is composed of layers of materials that differ in composition and density
4. Investigate and explain the various earth cycles including the rock cycle, coastal and ocean tides, and geological eras
5. Explain how the earth maintains its oxygen content in the air
6. Analyze and conclude how the slow movement of material within the earth results from heat flowing out from the deep interior and the action of gravitational forces on regions of different density

Science: **Performance**
Grade 3 **Benchmark**

EARTH SCIENCE: GEOLOGY
CONTENT/CONCEPT STANDARD 1

KEY ORGANIZING QUESTION:
How are environments similar and different?

KEY COMPETENCES	KEY CONCEPTS AND CONTENT	PERFORMANCE TASKS
Explore Describe Explain Illustrate Write Present Question	Variation: Surface features vary. Humans can create changes in the natural features.	**PERFORMANCE TASK I:** Explore the geological characteristics of the region where you live. Some of the features you see around you are natural, and others are constructed. Explain how you can tell if a geological feature was made by humans or naturally caused. Select one geological feature and explain how it is similar to land features in other regions. Illustrate your ideas and write a descriptive paragraph explaining your picture. Present your picture and paragraph to the class and then combine all of them into a book that can be shared with students in other classrooms. Ask them for feedback on your class book. **PERFORMANCE TASK II:** You are a travel agent and have been asked to create a "travelogue" for visitors to your region. Explore the geological characteristics of the region where you live. Through pictures, illustrations, and words describe the geological features that one would see there. Be sure to distinguish between natural and humanly constructed features. Identify other regions that share similar features. Present your finished travelogue to a local travel agent and ask for reaction to it.

QUALITY CRITERIA:
"LOOK FORS"

- Clearly identify your purpose.
- Select various geological features.
- Identify and compare important characteristics.
- Represent your ideas using appropriate materials.
- Organize information logically.
- State main ideas with supporting details.
- Practice presenting for an audience.
- Ask for feedback from your audience.

Science: **EARTH SCIENCE: GEOLOGY** Performance
Grade 5 CONTENT/CONCEPT STANDARD 1 Benchmark

KEY ORGANIZING QUESTION:
What kind of information can be obtained from soil samples?

KEY COMPETENCES	KEY CONCEPTS AND CONTENT	PERFORMANCE TASKS
Examine Identify Describe Develop Recommend Support Illustrate	Systems Patterns of Change Scale and Structure: Various levels of soil have different characteristics. Sediments provide an historical record of the conditions of a system. The substances found in sediments are often the only evidence for larger changes.	**PERFORMANCE TASK I:** A construction company is often required to take soil samples before construction begins. Examine and describe the soil samples and develop a list of things that a contractor would be interested in when studying a soil sample. Acting as an adviser to a construction company, write a recommendation to the contractor stating whether or not he or she should build on the site where the soil was collected. Support your conclusion with illustrations and related data. (Note to teacher: Sample of soil must be an intact clump from the surface to about a foot deep. It can be displayed in a large container.)

QUALITY CRITERIA:
"LOOK FORS"
- Clearly define your purpose.
- Accurately observe and identify the characteristics of sediments.
- Describe the unique characteristics of the sample.
- Create lists of significant details.
- Draw a conclusion based on your evidence and observation.
- Support your recommendation with data.
- Create an illustration to support your findings.

PERFORMANCE TASK II:
Collect an intact soil sample approximately 1 foot deep from your local neighborhood. Examine and describe this soil sample. Develop a list of characteristics about the sample. What information can you obtain from this sample? Write an article for the Sunday newspaper and recommend whether or not structures should exist on this site. Support your conclusion with illustrations and related data.

Science: Grade 8

Performance Benchmark

EARTH SCIENCE: GEOLOGY
CONTENT/CONCEPT STANDARD 2

KEY ORGANIZING QUESTION:
What are the effects of various natural forces on living and nonliving things?

KEY COMPETENCES	KEY CONCEPTS AND CONTENT	PERFORMANCE TASKS
Select Investigate Analyze Conclude Transfer Design Develop Draw Present Explain	Systems Patterns of Change Ecology: How natural forces affect living and non-living things Human adaptation to various natural forces	**PERFORMANCE TASK I:** Think of yourself as an architect who chooses to design buildings that are compatible with nature and the possible forces of nature. To ignite your thinking, select an animal; then investigate and analyze its natural habitat. Draw conclusions about the animal's habitat construction that provide evidence of its working with or against the environment. Select one or two major contributions and transfer them to the development of a sound ecological building design for people. Design and develop your drawing for an ecologically sound building for people. Present your illustration to a panel of architects and explain your innovative ideas and how you developed them. **PERFORMANCE TASK II:** As an ecologist, select a place or structure in your neighborhood that has suffered from the forces of nature, or could very easily become the target of a natural force. Investigate and analyze the conditions of your selection and conclude what changes would have to occur in order to reduce or eliminate destruction from the forces of nature. Select one or two of your major ideas and transfer them to a drawing that depicts the changes you are recommending. Carefully design and develop your drawing and then present and explain it to a representative of your local environmental protection agency or to students in another science class.

QUALITY CRITERIA:
"LOOK FORS"
- Clearly state the problem.
- Collect information from a variety of resources.
- Organize the collected data.
- Identify key conclusions.
- Relate your findings to the stated problem.
- Create a representation of your idea.
- Include necessary accurate details to convey your ideas.
- Review possibilities and edit according to need.
- Extend your ideas to the fullest visually.
- Prepare for your audience.
- Select and use appropriate language and style.
- Elicit feedback and respond accordingly.

Content/Concept Standards for Science

Science: Grade 12

Performance Benchmark

EARTH SCIENCE: GEOLOGY
CONTENT/CONCEPT STANDARD 1

KEY ORGANIZING QUESTION:

What is the relationship between human activity and the environment?

KEY COMPETENCES	KEY CONCEPTS AND CONTENT	PERFORMANCE TASKS
Gather Describe Analyze Propose Design Develop Present Explain	Systems Patterns of Change: Humans as part of and change agents of the environment. Systems adapt to changes in their environment.	**PERFORMANCE TASK I:** Gather information so you can describe the history of your city, including geological factors that caused people to settle there. Include an analysis of the natural resources that were originally available. Explain how the original resources have changed over time. Predict and categorize the resources that will be necessary for the continued existence of the city. Design and develop a blueprint of how your city might appear in the next 50 years based on your prediction of resources. Conclude why this information is important for current leaders and citizens. Present your findings, your recommendations, and your blueprint for the future to selected community representatives and explain your position. Be sure to include recommendations that will benefit the citizens, the environment, and the city. **PERFORMANCE TASK II:** Gather information on a region that has been abandoned for its lack of resources. Describe and analyze the human activities that, in part, caused the abandonment. Describe in a narrative how human activities have changed as a result of this "resource loss." Illustrate these relationships over a period of time in a visual presentation. Explore viable options for the development of this area. Select your best idea, and design and develop a plan to improve this region so it can become a thriving area. Present and explain your ideas to representatives from the business community.

QUALITY CRITERIA: "LOOK FORS"

- Clearly state your purpose.
- Collect and organize viable resources and data.
- Identify the key issues.
- Use various techniques to compare and contrast changes.
- Prioritize the gathered information.
- Anticipate options for the future based on your information.
- Make recommendations.
- Create a representation, including necessary details.
- Review and edit your representation.
- Arrange all necessary details.
- Create the final product.
- Choose an appropriate manner for disseminating to selected audience.
- Evaluate response from audience.
- Identify new possibilities.

EARTH SCIENCE – SPACE SCIENCE

Content/Concept Standards

Space science as the study of the Earth in relation to its surroundings requires us to learn about the composition of the universe and the nature of space and time, the principles governing the observed behavior of the components of the universe, and the historical developments that led to our present views and efforts in space science.

What students should know how to do by the end of Grade 3

Space science at this level should be observational and qualitative in nature. Students should, through short- and long-term observations, be able to describe what they see in the day and night sky. Observations of size, shape, brightness, numbers of objects, length of day and night, and time between phases of the moon should be recorded and discussed. Students should try to relate these observations to if and how they affect the Earth and people living on the earth. They should be able to

1. Examine and explain how and why our behavior changes as the seasons change
2. Identify the simple observable characteristics of the objects that we see in the day and night sky
3. Describe and record the observed, relative positions and motion of the moon and sun over a few hours, days, and months
4. Develop explanations for the motion of the earth, stars, moon, and sun

What students should know how to do by the end of Grade 5

Planets differ in their appearance and motion. The notion that the stars move across the sky differently than planets is to be discussed and alternative explanations evaluated. The space program has yielded considerable information, and videos will provide additional information for students to learn the conditions and nature of outer space. They should be able to

1. Develop an inventory of the variety of things in the universe
2. Evaluate the various geocentric and heliocentric models for planetary motion
3. Determine the conditions that support life on earth that must be provided to support life in space
4. Explore and observe differences in light sources
5. Research and discuss the nature and conditions of outer space
6. Use telescopes to observe distant objects in the sky

What students should know how to do by the end of Grade 8

This is a time for students to begin to learn the complexities of space science through the use of technology and the construction of models based on their observations and measurements. The notion of scale and how distances are measured should be introduced. Various resources are available to investigate the importance of space studies and the difficulties involved. Students should be able to

1. Research and analyze the benefits and problems of space travel as a necessary scientific endeavor
2. Investigate and compare the relative motions of the moon, stars, and planets
3. Explore size and distance through the construction of models
4. Estimate distances by triangulation and scale drawings
5. Explore the concept and the content of galaxies and their numbers

What students should know how to do by the end of Grade 12

Students should have the evidence to piece together a sound concept or mental model of the universe. This evidence should support conclusions that many things are known about the properties of space and the objects in space from our earthbound vantage point, including the scale and the age of the universe, our solar system, and the earth. Students should have the evidence that supports the conclusion that such things as the universal law of gravitation and the nature of light are invariant for space-time. Further, various commercial ventures are planned in space that include communication, pharmaceuticals, and metallurgy. Students must understand why these ventures are important to people on Earth. They should be able to

1. Determine how we can learn if the principles of science are the same in space as they are on the Earth
2. Research the benefits that have been achieved from our space explorations and predict how we may benefit from further investigations
3. Examine the matter of scale when comparing space travel to travel on earth
4. Demonstrate and contrast how human activities affect land, sea, air, and space
5. Explore the uses of various technologies in the acquisition of information about the universe

Science: **Performance**
Grade 3 **Benchmark**

EARTH SCIENCE: SPACE SCIENCE
CONTENT/CONCEPT STANDARD 2

KEY ORGANIZING QUESTION:
What can you learn by observing the day and night sky?

KEY COMPETENCES	KEY CONCEPTS AND CONTENT	PERFORMANCE TASKS
Review Create Predict Observe Identify Record Explain Compare	Patterns of Change: Objects will vary in properties from brightness to size and shape. Observe and discuss the different characteristics of the objects found in the day and night sky.	**PERFORMANCE TASK I:** Review different information on which objects can be found in the sky. Create a list of objects you might be able to observe in the night sky and in the day sky without the aid of special instruments such as telescopes. Discuss how you might tell the difference between stars and planets. Observe the day sky and the night sky carefully. Record your observations in two sketches: "Day Sky" and "Night Sky." Present and explain your observations to a learning buddy. Compare the similarities and differences that you each observed and recorded. Display your sketches for others to view. **PERFORMANCE TASK II:** Review various information about the moon. Create a list of all the things you might see by observing the moon in the night sky. Predict some of the changes you might see. Observe the moon in the night sky every night at the same time for 2 or 3 weeks. Carefully record your observations in a sketch. Be sure to date your sketches. Present and explain your observations to a learning buddy. Compare the similarities and differences that you each observed and sketched. Display your sketches for others to view.

QUALITY CRITERIA:
"LOOK FORS"
- Clearly state your purpose.
- Utilize a variety of resources.
- Represent ideas in a list.
- Select and state appropriate setting for observations.
- Create clear, neat drawings of your observations.
- Include date and time on your sketches.
- Explain similarities in drawings.
- Explain differences in drawings.

Content/Concept Standards for Science

Science: Grade 5

Performance Benchmark

EARTH SCIENCE: SPACE SCIENCE
CONTENT/CONCEPT STANDARD 2

KEY ORGANIZING QUESTION:
How can we learn about planetary motion?

KEY COMPETENCES	KEY CONCEPTS AND CONTENT	PERFORMANCE TASKS
Review Design Illustrate Discuss Write Present Explain	Systems Patterns of Change: The planetary system is bound by gravitational forces. Planetary motion is very predictable and can be successfully explained by different models.	**PERFORMANCE TASK I:** Review various historic and present models of the planetary system to which the Earth belongs. After discussing the unique features, design and illustrate a model of a planetary system that includes two stars and five planets. Discuss how the two-star system would move. Write a descriptive paragraph about your illustration. Check to see if your illustration accurately represents motions of objects in the sky. Present your illustration to a group of 4th-grade science students. Explain how the various objects move in the sky. **PERFORMANCE TASK II:** Design your own planetary system. It can be a single-star or a two-star system. Write an explanation for the system in your drawing. Select students in your class to represent the various objects in your planetary system. Direct their movements so they move as objects in the sky. Present your live demonstration to another class and carefully explain the motions of the objects in your planetary system.

QUALITY CRITERIA:
"LOOK FORS"
- Identify your purpose.
- Utilize a variety of resources.
- Document your primary information.
- Share your information with another learner.
- Explain your ideas on movement of objects.
- Create a detailed illustration of the planetary system.
- Organize your ideas and materials.
- Present your information to a live audience.

Science: Grade 8

Performance Benchmark

EARTH SCIENCE: SPACE SCIENCE
CONTENT/CONCEPT STANDARD 1

KEY ORGANIZING QUESTION:
Is space travel a necessary endeavor?

KEY COMPETENCES	KEY CONCEPTS AND CONTENT	PERFORMANCE TASKS
Compare and Contrast Identify Describe Defend Write Submit	Systems: Economic and social systems are linked in many ways. Space travel as a scientific endeavor for humankind.	**PERFORMANCE TASK I:** Compare and contrast the benefits and problems of having and supporting a space science program. Identify the direct benefits of the space program and compare them to the benefits we might get if the money from that program were allocated to reducing the deficit. What if the space program were stopped? Categorize short and long term effects of cancelling the space program. How would this affect the United States? Publish your conclusions in a news article (or space team newsletter) to raise public awareness. Submit your article to an authentic publisher. **PERFORMANCE TASK II:** Compare and contrast the benefits and problems of having and supporting a collaborative endeavor like the Mir Space Station. Identify the direct benefits to our society and compare then to the benefits we might get if the money from that program were allocated to reducing the deficit. What impact might there be if work on the space station was canceled? What impact might result if funds and involvement in the space station were increased? Publish your conclusions in a news article (or space team newsletter) to raise public awareness. What else might you do to increase awareness of your position?

QUALITY CRITERIA: "LOOK FORS"
- Identify all important components.
- Use variety of techniques to compare and contrast information.
- Select the most important ideas to address.
- State points clearly.
- Provide necessary supportive details and examples.
- Create a draft for the magazine article.
- Edit the draft and create final copy.
- Submit to authentic publication.

Content/Concept Standards for Science

Science: Grade 12

Performance Benchmark

EARTH SCIENCE: SPACE SCIENCE
CONTENT/CONCEPT STANDARD 1

KEY ORGANIZING QUESTION:
How can we learn if the laws of science are universal and constant over time?

KEY COMPETENCES	KEY CONCEPTS AND CONTENT	PERFORMANCE TASKS
Reflect Identify Support Describe Write Present Defend Assess Submit	Patterns of Change: We can predict the behavior of distant objects by applying scientific principles developed on earth.	**PERFORMANCE TASK I:** You are a scientific investigator preparing to present a paper for a scientific convention on universal laws. You are proposing that there are certain "universal" characteristics for all the planets. Because this is a very controversial stand to be taking, you will need to support your presentation with convincing visuals and factual information. Identify possible challenging questions you might get from your audience and develop strong responses. Present your paper to another class and encourage their reactions and arguments. Reflect on your ideas and your position. Create a final paper and use it as the basis for submitting a proposal to a professional scientific society conference. **PERFORMANCE TASK II:** You are a scientific investigator. Review the possibilities of a plan to test Newton's laws of motion regarding the relationship of objects in space with earth. Write a plan to test these possibilities. Include persuasive information about shared characteristics of earth and other planets. Based on this plan, develop a speech with supporting visuals to convince a group of experts that you have proof of "universal laws." Present your speech to members of your class and encourage their reactions and arguments. Reflect on your ideas and your original plan. Create a final paper and use it as a basis for submitting a proposal to a professional science society conference.

QUALITY CRITERIA:
"LOOK FORS"
- Reflect on key scientific principles related to the universe.
- Clearly identify your position and purpose.
- Review a variety of resources.
- Identify beginnings.
- Arrange details.
- Expand the concept.
- Draft and edit a document.
- Review with colleagues.
- Submit a proposal to present to a professional organization.

2
TECHNOLOGY CONNECTIONS

SUMMARY

Why Address Technology in a Performance-Based Curriculum?

A performance-based curriculum starts with the understanding that students will make use of what they learn in the production and dissemination of knowledge. Technology is revolutionizing the way we access information; the capabilities we have in interpreting and analyzing data; the methods by which we produce, design, and construct products resulting from our learning; the forms those products take; the methods by which the products are disseminated; and the evaluation procedures we can undertake. *Access, interpret, produce, disseminate,* and *evaluate:* These are the five central learning actions in a performance-based curriculum. These learning actions used in conjunction with technology give the learner more power and lead to greater effectiveness.

PERFORMANCE-BASED LEARNING ACTIONS WHEEL

Technology as Content

Our physical, social, and material worlds are being radically changed as a result of the explosion of new technologies. Technological change and the issues stemming from that change provide content that is increasingly addressed in the study of history, economics, political science, and other disciplines making up the social sciences. They are also subject matter for novels, science fiction, and political and social essays. Technology is a central focus of futuristic studies. It is a product of, as well as a critical ingredient in, modern science. Technological developments have radically altered the tools used by authors and everyone involved in communication and the use of language. Technology is a rich source of topics for integrating a performance-based curriculum.

Technology as a Tool

Technology is also used as a tool in a performance-based curriculum. Although technology can be used as a way of controlling the learner's interaction with the curriculum, technology is most appropriately used as a tool controlled by the learner in the performance-based approach to learning. It is that approach that is applied in correlating this section with the Science section.

Many technologies can enhance a performance-based curriculum. Their common characteristic is that they are tools that improve communication of and access to multimedia data (words, numbers, sounds, still and motion pictures, still and motion graphics) and make the use of those data easier and more effective. In a perfect world, every student and teacher would have a workstation equipped with a computer, modem, CD-ROM, laserdisc player, and a videotape camera and player. This workstation would be connected to networks that allow access to multimedia data on demand. The networks would distribute information in multimedia format to others throughout the world. In addition to these workstations, teachers and learners would have access to copying, scanning, and printing machines; CD-ROM presses; video editing equipment; audio recording and editing equipment; and software to support writing, computer-aided design, statistics, graphing, musical and artistic productions, and so on. Additional equipment would be found in a science laboratory, including tools for specialized data collection and analysis. In other specialty areas, such as art, lithographic presses would be available. Drafting equipment, electronic tools, and other specialized technologies would be present where necessary to allow the teaching of those technological subject areas.

Technology is a tool (among other tools) useful for acquiring, storing, manipulating, and communicating information in a multimedia format. Technology can be used to gather data, explore questions, produce products, and communicate results.

Technology in Support of Learning Actions

Five learning actions are central to a performance-based curriculum: **ACCESS, INTERPRET, PRODUCE, DISSEMINATE**, and **EVALUATE**. Throughout this curriculum framework, the use of appropriate technologies will support students in being active learners. Students will be encouraged to use technology to generate questions and identify problems in a wide variety of contexts; formulate hypotheses and generate tentative solutions to the questions or the problems they have defined; test the reasonableness of their answers and respond to challenges to their positions; reach a conclusion about an issue, a problem, or a question and use that "solution" as a jumping-off place to ask other questions; and engage in the learning process again.

A learner with a purpose, an issue, a question, or an idea needs to be able to use appropriate technologies in carrying out these learning actions. Technology is especially important in accessing information, producing products, and disseminating the results of one's work. We organize the benchmarks of the skills students must have in using technology around these key learning actions that can take full advantage of current technologies: **ACCESS, PRODUCE**, and **DISSEMINATE**. Examples have been developed for two strands of each of the grade levels in mathematics. Each example contains suggestions on how to use technology to **ACCESS** information, **PRODUCE** products, and **DISSEMINATE** the results of one's efforts. These examples are meant to stimulate and facilitate the mastery of the use of appropriate technologies in the pursuit of learning. The suggested technologies encompass a broad range of tools useful in accessing, producing, and disseminating data that are not just words and numbers but are also sounds, still and motion graphics, and still and motion pictures. Students and teachers are encouraged to use all appropriate tools and disseminate their products using a combination of technologies.

Technology changes rapidly. The skills and abilities described below require modification on a regular basis to reflect the latest technologies. These skills and abilities must be understood as dynamic objectives rather than as static goals. They are essential learning actions that increase the student's ability to **ACCESS, PRODUCE**, and **DISSEMINATE**.

Technology Connections for Science

SKILLS AND ABILITIES

How students should be able to use technology by the end of Grade 3

Access:

A1	Gather information with still, digital, or video camera
A2	Search databases to locate information
A3	Gather sounds and conversations with audio and video recorders
A4	Collect digitized audio data
A5	Access information on laserdisc by using bar code reader
A6	Scan to capture graphic data
A7	Copy to gather graphics
A8	Retrieve and print information using a computer
A9	Gather information through telephone
A10	Select and use information from CDs
A11	Fax to send and receive printed information
A12	Identify and use all types of materials, such as print, nonprint, and electronic media
A13	Locate information using electronic indexes or media

Produce:

P1	Draw and paint graphics and pictures using a computer
P2	Create flip card animations using a computer
P3	Design and develop computer products including pictures, text, flip card animations, sounds, and graphics
P4	Design and develop audiotapes
P5	Design and develop videotapes
P6	Create overhead or slide presentations with or without background music
P7	Develop stories using computer-generated text with either handmade or computer-generated illustrations

Disseminate:

D1	Present *Logo* or *HyperCard* (or similar) computer product including pictures, text, flip card animations, sounds, and graphics
D2	Publish printed page including text and graphics
D3	Broadcast audiotape
D4	Broadcast videotape
D5	Present overhead or slide presentation
D6	Fax information to other audiences
D7	Explain products or creations to an audience

How students should be able to use technology by the end of Grade 5

Access:

A1 Gather information with a still, digital, or video camera of moderate complexity
A2 Gather information using text-based databases to locate information
A3 Access information on laserdisc by using bar code reader and computer controls
A4 Gather information using telephone and modem to connect to other users and databases (Internet, eWorld, etc.)
A5 Search basic library technologies for data
A6 Select and use specialized tools appropriate to grade level and subject matter
A7 Record interviews with experts
A8 Scan CD collections for needed information

Produce:

P1 Create path-based animations using computer
P2 Create with computer painting and drawing tools of moderate complexity
P3 Digitize still and motion pictures
P4 Create basic spreadsheet for addition, subtraction, multiplication, and division
P5 Graph data (pie charts, line and bar graphs) using computer
P6 Create edited videotapes of moderate complexity using a videotape editing deck or computer-based digital editing system or two connected cassette recorders (VCRs)
P7 Input text into computer using keyboard with appropriate keyboard skills
P8 Design and develop moderately complex *Logo* or *HyperCard* (or similar) programs including pictures, sounds, flip card and path-based animations, graphics, text, and motion pictures
P9 Design and develop multipage document including text and graphics using computer
P10 Create edited audiotape
P11 Create edited videotape
P12 Create overhead or slide presentation with synchronized voice narration with or without background music
P13 Lay out advertisements, posters, and banners

Disseminate:

D1 Present moderately complex *Logo* or *HyperCard* (or similar) computer product including pictures, sounds, flip card and path-based animations, graphics, text, and motion pictures
D2 Publish multipage printed document including formatted, paginated text and graphics
D3 Broadcast edited audiotape and videotape
D4 Present programs using overhead projector, slide projector, or computer
D5 Present information over public address system in a school, community, or meeting situation
D6 Display information in a variety of formats
D7 Advertise for events, services, or products
D8 Broadcast performances and products
D9 Broadcast on cable TV

Technology Connections for Science

How students should be able to use technology by the end of Grade 8

Access:

- A1 Gather information using computer, CD-ROM, and laserdisc databases
- A2 Gather data using telephone and modem (including graphics and sounds) to and from other users and databases (Internet, eWorld, etc.)
- A3 Search basic spreadsheet and databasing software for "what if?" comparisons and analyses
- A4 Search technologies for accessing data outside the school and local library
- A5 Search menus to locate information on computer software, CD-ROM, or laserdiscs
- A6 Video interviews
- A7 Download information from Internet

Produce:

- P1 Create products using computer painting and drawing tools, including moderately complex color tools
- P2 Digitize still and motion pictures
- P3 Create edited videotapes by using a videotape editing deck or computer-based digital editing system
- P4 Create computer presentation program
- P5 Develop cell-based animations using computer
- P6 Design and develop complex *Logo* or *HyperCard* (or similar) programs including still pictures; flip card, path-based, and cell-based animations; sounds; graphics; and motion pictures
- P7 Create multipage documents including text and graphics using computer page layout tools
- P8 Develop audiotapes that combine sounds and voice data from a variety of sources
- P9 Produce videotapes that are organized, coherent, and well edited
- P10 Create a personal database requiring the collection of data over time

Disseminate:

- D1 Present relatively complex *Logo* or *HyperCard* (or similar) product including still pictures; flip card, path-based, and cell-based animations; sounds; graphics; and motion pictures
- D2 Publish multipage printed documents including text and graphics
- D3 Broadcast edited audiotape of moderate complexity
- D4 Broadcast edited videotape of moderate complexity
- D5 Broadcast video presentation over schoolwide Channel 1 (Whittle), citywide public Channel 28, or citywide ITFS schools-only equipment
- D6 Advertise events, services, or products
- D7 Display information and designs on various formats available
- D8 Broadcast on closed circuit or cable television
- D9 Broadcast filmed and live performances on television
- D10 Distribute over available sources in Internet

How students should be able to use technology by the end of Grade 12

Access:

A1	Access and use complex electronic databases and communication networks of all types including, but not limited to, Internet
A2	Research using sensors, probes, and other specialized scientific tools as appropriate
A3	Gather information from spreadsheet, databasing software, and statistical packages, including the use of formulas and charting routines
A4	Search technologies for data and primary sources (publications and persons)
A5	Identify local, regional, and national databases and procedures for needed data
A6	Review online bulletin boards, databases, and electronic retrieval services for data

Produce:

P1	Create with complex computer painting and drawing tools and programs
P2	Create 3-D graphics using drawing and modeling tools
P3	Create changing images using computer digital-morphing programs
P4	Illustrate concrete and abstract concepts using computer-aided design and mathematical modeling
P5	Create CD-ROM simulations
P6	Create complex cell-based animations, including 3-D objects, using the computer
P7	Create complex *Logo* or *HyperCard* (or similar) programs including pictures; flip card, path-based, and cell-based animations; sounds; 3-D graphics; and motion pictures
P8	Develop multipage documents with information from a variety of sources, including text and graphics using appropriate computer page layout tools
P9	Create documents using a variety of fonts and type faces
P10	Assemble findings based on spreadsheets, databasing software, and statistical packages involving the use of formulas as appropriate
P11	Design graphic and text titles for digital video productions
P12	Develop digitally edited materials including audio, motion pictures, still-frame pictures, motion graphics, and still-frame graphics
P13	Design and develop a personal database of moderate complexity
P14	Illustrate concrete and abstract mathematical and scientific concepts
P15	Assemble information by creating, searching, and sorting databases
P16	Design and develop a dissemination design for video using ITFS microwave and satellite up-and-down links

Disseminate:

D1 Transmit complex *Logo* or *HyperCard* (or similar) computer product including pictures; flip card, path-based, and cell-based animations; sounds; 3-D graphics; and motion pictures

D2 Publish multipage printed documents, appropriately laid out, including text and graphics

D3 Transmit complex spreadsheet or database findings

D4 Telecast digital video product of some complexity

D5 Present computer-based animation program (cell- or path-based animations, or both)

D6 Publish reports generated from database searches

D7 Publish scientific investigations and results or recommendations

D8 Transmit a video presentation to secondary students using ITFS microwave, Whittle Channel 1 equipment, public Channel 28, cable hookups, and satellite up-and-down links to local schools or students in other school systems

Technology Connections
Science: Grade 3

Performance Benchmark

PHYSICAL SCIENCE: MATTER
CORRESPONDING PERFORMANCE BENCHMARK, PAGE 29

KEY ORGANIZING QUESTION:
How and why are common measuring instruments used?

ACCESS	PRODUCE	DISSEMINATE
PERFORMANCE TASK I: Using your ruler, balance, and thermometer, determine the lengths and weights of the objects identified for exploration in your room. Also determine the temperature of the place where the object is located at the time you take your length and weight measurements.	**PERFORMANCE TASK I:** Using your word processor, create a spread sheet. Make a list of all the objects you weighed and measured. Make a column for the weight of the object, the length of the object, and the temperature of the place where the object is located and the time, and add the data you collected to the appropriate column.	**PERFORMANCE TASK I:** Print your spreadsheet. Now create a graph using a spreadsheet, database, or statistics package, if available, or by hand. Can you see any relationships between the temperature and the weight? Between length and weight? Between temperature and length? Describe your conclusions to another student or team.
PERFORMANCE TASK II: Find five heavy objects in your home or room at school. Find five light objects in your home or room at school. Measure the length of each.	**PERFORMANCE TASK II:** On your word processor, make a list of the five heavy objects and the five light objects. In a column next to the name of the object, record its length. Are any of the heavy items shorter than the longer items? If so, why do you think that is the case? If all of your heavy objects are longer than your light objects, can you find a light object that is longer than at least one of your heavy objects? Add it to your list. Make a list, also on your word processor, of the reasons why you think there is, or is not, a relationship between length and weight. Also write down your answer to the question, "As you get taller, will you get heavier?" Explain your answer.	**PERFORMANCE TASK II:** Print and graph your list of objects and their weights, your reasons for believing there is or is not a relationship between length and weight, and your answer and explanation to the question, "As you get taller, will you get heavier?" Justify your conclusions in an explanation to a classmate.

Technology Connections
Science: Grade 5

Performance Benchmark

PHYSICAL SCIENCE: MATTER
CORRESPONDING PERFORMANCE BENCHMARK, PAGE 30

KEY ORGANIZING QUESTION:

How are chemical and physical changes differentiated?

ACCESS	PRODUCE	DISSEMINATE
PERFORMANCE TASK I: Using your word processor, make a list of the demonstrations conducted by your teacher. Indicate in a separate column whether the change that takes place is a chemical change or a physical change. At the bottom of your list, write down the definitions that you develop for physical change and chemical change.	**PERFORMANCE TASK I:** List all of the possible combinations that could be made of these items using two at a time: weak vinegar, baking soda, salt, sugar cubes, water. Before conducting your experiments, guess whether combining the two substances or items will produce a chemical reaction or a physical reaction. List your prediction in a column by each pair. Carry out your experiments. In another column, list whether the results of the experiment indicate the combination produced a chemical reaction or a physical reaction. Print your report.	**PERFORMANCE TASK I:** Present your findings orally. Explain your results to another student or team. Discuss any differences between your findings and those of other groups.
PERFORMANCE TASK II: Using a thermometer, determine the temperature of salt, water, and ice chips. Record these data using a computer in a word processing program, database, or spreadsheet. Mix these ingredients together by pairs. Determine their temperature and again record the data.	**PERFORMANCE TASK II:** Using your word processor, make a list of these three items and each of the paired combinations. In separate columns, describe the look of each, the feel of each, and the temperature of each. In a fourth column, assign a category to each substance or mixture using the definition created by your group to classify changes. Prepare an ad to sell salt for use on frozen sidewalks. Use your findings as part of an explanation as to why they should buy the salt.	**PERFORMANCE TASK II:** Distribute your ad to an appropriate audience on the Internet.

Technology Connections
Science: Grade 8

Performance Benchmark

PHYSICAL SCIENCE: MATTER
CORRESPONDING PERFORMANCE BENCHMARK, PAGE 31

KEY ORGANIZING QUESTION:
How can we use models to explain correctly how physical changes occur from an atomic perspective?

ACCESS	PRODUCE	DISSEMINATE
PERFORMANCE TASK I: 1. What happens in the cleaving of a crystal? Describe this in words or drawings by using your word processor or the draw or paint tools on your computer. 2. What happens when a gas expands? Describe this in words or drawings by using your word processor or the draw or paint tools on your computer. 3. What happens when sugar is dissolved in water? Describe this in words or drawings by using your word processor or the draw or paint tools on your computer.	**PERFORMANCE TASK I:** 1. Using flip card animation tools—*HyperCard*, *Logo*, or another appropriate program—and the drawing or paint tools of your computer, create a flip card animation of the cleaving of a crystal that fits the explanation you have written down. 2. Using flip card animation tools—*HyperCard*, *Logo*, or another appropriate program—and the drawing or paint tools of your computer, create a flip card animation of the expansion of a gas that fits the explanation you have written down. 3. Using flip card animation tools—*HyperCard*, *Logo*, or another appropriate program—and the drawing or paint tools of your computer, create a flip card animation of the dissolving of sugar in water that fits the explanation you have written down.	**PERFORMANCE TASK I:** Show your flip card animations to someone who does not understand what happens in the physical changes you are examining. After showing your animation to him or her, ask questions to see how much he or she now understands.
PERFORMANCE TASK II: Locate examples of models that represent good visuals of physical changes in a CD or digital database, encyclopedia, or dictionary.	**PERFORMANCE TASK II:** Using your word processor, create your own verbal analogy for one of the changes you have represented with a graphic.	**PERFORMANCE TASK II:** Print your analogy and your visual graphic and share it with others. Ask questions to determine whether your analogy helps them to understand the changes that you are trying to describe by analogy.

Technology Connections for Science

Technology Connections Science: Grade 12

Performance Benchmark

PHYSICAL SCIENCE: MATTER
CORRESPONDING PERFORMANCE BENCHMARK, PAGE 32

KEY ORGANIZING QUESTION:
How can events in our daily lives be related to particles of the atom?

ACCESS	PRODUCE	DISSEMINATE
PERFORMANCE TASK I: Gather information from a CD-ROM database or the Internet on Ionic, covalent, and hydrogen bonds.	**PERFORMANCE TASK I:** Design and develop graphics to show examples of the three types of bonding. Include a description of the graphics explaining the physical characteristics associated with each bond type. Be sure to include any additional information you discovered related to other compounds. Use these graphics to create a computer slide show depicting examples from daily living that explain the various bondings.	**PERFORMANCE TASK I:** Present your computer slide show to your class. Donate the slide show, which could serve as a resource for students in future classes, to a class library.
PERFORMANCE TASK II: Gather information from a CD-ROM database, a digital dictionary, or a digital database on ionic bonds, covalent bonds, and hydrogen bonds. Determine which characteristics might be associated with each bond type. Test a variety of safe compounds, then do a comparison of the physical characteristics you observed and those you discovered in the database.	**PERFORMANCE TASK II:** Design and develop a graphic design to depict the three types of bonding to be included in a textbook to illustrate the concept. Include a description of your design explaining fully the characteristics associated with each bond type. Be sure to include any additional compounds found for the various bond types!	**PERFORMANCE TASK II:** Publish your text contribution. Share it with classmates. Ask for their reactions and recommendations. Edit accordingly, and make the final copy available to the teacher to use with students in future classes.

Technology Connections
Science: Grade 3

Performance Benchmark

EARTH SCIENCE: GEOLOGY
CORRESPONDING PERFORMANCE BENCHMARK, PAGE 46

KEY ORGANIZING QUESTION:
How are environments similar and different?

ACCESS	PRODUCE	DISSEMINATE
PERFORMANCE TASK I: The first task will be to collect pictures of geological features of your environment, such as rocks, rivers, hills, mountains, lakes, cliffs, fields, caves, and so forth. You can get the pictures from magazines, books, CD-ROMs, or from pictures available to you on your computer, or you can take pictures with a simple camera or a digital camera. Whatever the source of your pictures, scan them or digitize them into your computer, or paste them into your paint or drawing program or your word processor.	**PERFORMANCE TASK I:** Place one picture on each page of your word processor or *HyperCard* card. Next to the picture, list how humans have changed the geological features shown in each of your pictures. Also list what appears not to have been changed by humans. Finally, list why you think things have changed or have not changed.	**PERFORMANCE TASK I:** Either print your documents or run them on your computer so that others can see what you have produced. Present your product to members of another class. Explain your findings and be prepared to answer their questions.
PERFORMANCE TASK II: Make a list of the places that will be in your "travelogue" that will encourage people to drive across the country. Using the Internet, America Online, or another electronic network, contact people who could provide you with information for your travelogue. In your travelogue, describe the clues that would reveal to the traveler the special aspects of the places you have chosen to include. For example, if you have a desert, a mountain area, and a beach, describe how the traveler would know when he or she was near or in that place. Include pictures in your travelogue that you have scanned (and maybe even taken), or that you have acquired from a CD database, a laserdisc, or from some other source. Be sure to identify the source of your pictures.	**PERFORMANCE TASK II:** Arrange the material you have collected, scanned, or otherwise gathered for your travelogue using your word processor or other computer programs available to you. Carefully create written descriptions for each of your pictures. Make your travelogue attractive and interesting so that you will be able to get others to use it.	**PERFORMANCE TASK II:** Print your travelogue. Distribute it to members of your class. Also send copies of your travelogue over the Internet, or the electronic network you used to acquire the information, to those people who sent you materials for your travelogue (whether you used the materials or not).

Technology Connections for Science

Technology Connections Science: Grade 5

Performance Benchmark

EARTH SCIENCE: GEOLOGY
CORRESPONDING PERFORMANCE BENCHMARK, PAGE 47

KEY ORGANIZING QUESTION:
What kind of information can be obtained from soil samples?

ACCESS	PRODUCE	DISSEMINATE
PERFORMANCE TASK I: Locate in a digital database, a digital encyclopedia, a CD-ROM, or a laserdisc several pictures of different kinds of soil samples.	**PERFORMANCE TASK I:** Copy or digitize the picture or pictures of your soil samples into your drawing or paint program, your word processor, or *HyperCard*. Your pictures will also have material explaining the various features found in the soil samples. On a separate card or page, make a list of the places from which the samples in your picture were taken. In a separate column, list the various characteristics of the soil. In a third column, list your explanation for the similarities and differences that appear in your various pictures.	**PERFORMANCE TASK I:** Print your product and share with others, or use your computer to present the creation to your audience.
PERFORMANCE TASK II: Locate on a digital database, a digital encyclopedia, a CD-ROM, or a laserdisc a picture of a soil sample from a beach.	**PERFORMANCE TASK II:** Develop a list of features that fully describe the soil in your picture. Put these features into a column. In a separate column, list what each of these features might tell you about the history of the beach.	**PERFORMANCE TASK II:** Print your picture and descriptive columns. Present it to your classmates and explain fully. If you know someone who lives on or near a beach, fax them a copy of your printout.

69

Technology Connections
Science: Grade 8

Performance Benchmark

EARTH SCIENCE: GEOLOGY
CORRESPONDING PERFORMANCE BENCHMARK, PAGE 48

KEY ORGANIZING QUESTION:
What are the effects of various natural forces on living and nonliving things?

ACCESS	PRODUCE	DISSEMINATE
PERFORMANCE TASK I: The first task will be to collect pictures of animals that live in your environment. Each team member will be responsible for a different animal. You can get the pictures from magazines, books, or CD-ROMs, or from pictures available to you on your computer, or you can take pictures with a simple camera or a digital camera. Whatever the source of your pictures, scan or digitize them into your paint or drawing program or your word processor. In a digital encyclopedia, a CD-ROM, or other digital database, collect information on each animal about how that animal affects the environment (living and nonliving things). Pay special attention to the effects (if any) that each animal has on other animals on your list.	**PERFORMANCE TASK I:** In *HyperCard* or on your word processing program, create a document with one page or card for each animal. The picture should be on the card or page, as well as a list of the ways the animal affects the environment. In addition, create one card or page that shows the interactions among the animals and various aspects of the environment. One way to do this is with small pictures connected by arrows that show the direction of impact. There are other ways you might want to design your "summary."	**PERFORMANCE TASK I:** Print your project or run it for others on your computer. Arrange to share your project with fourth grade science students. Be sure to explain how you developed your computer presentation.
PERFORMANCE TASK II: Locate something in your community that has been adversely affected by the forces of nature. Take a picture of it and digitize it into your computer. You could also contact over a computer network someone in a place that has recently suffered a major natural disaster and get the person you contact to send you over the network or a fax machine, a picture or pictures of something that has been adversely affected by the natural disaster.	**PERFORMANCE TASK II:** Produce a plan to reduce the impact of the environment on the topic identified in your picture or the picture sent to you. Draw a picture (using the draw or paint programs of your computer) showing how your plan would help protect the thing in question. Write an essay describing your plan in words.	**PERFORMANCE TASK II:** Print your project or run it for others on your computer. Send a copy of your project over the network or fax it to the person who assisted you by supplying the picture or pictures.

Technology Connections for Science

Technology Connections
Science: Grade 12

Performance Benchmark

EARTH SCIENCE: GEOLOGY
CORRESPONDING PERFORMANCE BENCHMARK, PAGE 49

KEY ORGANIZING QUESTION:

What is the relationship between human activity and the environment?

ACCESS	PRODUCE	DISSEMINATE
PERFORMANCE TASK I: Locate pictures or maps of your community over the time of its existence. These pictures might be found in your local library or among the possessions of your parents, grandparents, neighbors, or friends. Businesses may have pictures of earlier times. The local government planning department might have copies available. Find maps that cover all or some of your community.	**PERFORMANCE TASK I:** Scan the maps and pictures you have found into your computer. Create a flip card animation or presentation using overheads that show the changes that have taken place by first displaying the oldest map, then the next oldest, and so on. You may have to adjust the size of your maps to a common size. Do this on your computer or copy machine. If you concentrate on pictures, arrange them by time also. Show the same block in town, the same corner, the same building, the same lot over time. Create an attractive program that will engage your audience, using the appropriate tools on your computer. Separately, make a list of the factors that explain the changes you have shown in your sets of maps and pictures. Also, make another list predicting changes that are likely in the future and why those changes will probably occur.	**PERFORMANCE TASK I:** Present your flip chart creation to a local community meeting, a realtors' association meeting, an art class, or a history class. Explain your discoveries and answer questions about the technical development of your creation.
PERFORMANCE TASK II: Take pictures of something in your community that has been abandoned, or is in the process of being abandoned. If you can, locate pictures taken before the abandonment took place.	**PERFORMANCE TASK II:** Scan the picture or pictures you have taken into your computer. If you have "before-and-after" pictures, arrange them appropriately. Make a list of the factors that led to the abandonment or the move toward abandonment. Make a list of what might be done to reverse the process of abandonment or improve the area, building, or focus in question. Create visuals that reflect the factors that may have contributed to abandonment and the factors or recommendations for improving the area, building, or topic of your focus.	**PERFORMANCE TASK II:** Make a technical presentation using a computer to your local city council or zoning committee explaining the relationship between human activities and the condition of the environment.

3
PERFORMANCE DESIGNERS

The ultimate key to success with performance-based education is the creativity, rigor, and consistency of focus that must characterize the ongoing instructional process in the classroom. Student success with the performance benchmarks identified in this text depends on daily interactions with the learning actions. Students must feel empowered to demonstrate the learning actions being taught so they can internalize them, take ownership, and apply them easily in the benchmark performances. They must be able to do this through a continuous improvement process with a focus on quality criteria.

In order to accomplish the performance benchmarks in this text, learners must have daily practice with the routine of learning and demonstrating through learning actions as they gain new understanding about concepts from the different disciplines. They must recognize that only through continuous improvement will they achieve the defined quality that must be their goal.

If this is to occur, teachers must design lessons specifically addressing the learning actions (access, interpret, produce, disseminate, and evaluate). Instruction on these learning actions will engage students in gathering and interpreting information so they can produce a product, service, or performance with their newly acquired insights and knowledge. Then they can disseminate or give their product, service, or performance to an authentic audience. They do all of these learning actions with a continuous focus on evaluating themselves and their work against the identified quality criteria that the teacher will be looking for.

The performance designer is a tool for teachers to use when planning for students to engage in a significant demonstration that is an interactive experience for students designed to include essential content, competence (learning actions), context (issue, situation, and audience), and quality criteria.

The completed performance designer will describe the total performance or demonstration of significance. All of the students' actions will be clearly stated. The teacher uses this performance designer to develop the necessary instructional sequences that will support the attainment of each of the desired actions. Once students know how to do the actions, they are ready to pursue the planned performance.

The following organizer provides an overview. Each major section in the planner is identified and corresponds to a detailed explanation that follows.

PERFORMANCE DESIGNER FORMAT

I
- **Ⓐ PURPOSE** .... What complex thinking process is the focus?
- **Ⓑ KEY ORGANIZING QUESTION** ...... An issue or challenge to investigate.
- **Ⓒ ROLE** .... You are _____ who is expected to ...

II
(Do what?)	(With what?)	(How well?)
Ⓓ Access and Ⓔ interpret by...	Ⓕ CONTENT/CONCEPTS	Ⓖ QUALITY CRITERIA "Look fors"

III
(In order to...)	(...do what?)	(How well?)
Ⓗ Produce by...	Ⓘ PRODUCT/ PERFORMANCE	Ⓙ QUALITY CRITERIA "Look fors"

IV
	(To/for whom? Where?)	(How well?)
Ⓚ Disseminate by...	Ⓛ AUDIENCE/ SETTING	Ⓜ QUALITY CRITERIA "Look fors"

Section I

The first section of the designer serves as an organizer for the actions that follow.

PERFORMANCE DESIGNER ELEMENT	REFLECTIVE QUESTIONS
❶ PURPOSE The reason the performance is worth doing. This section may be tied to state- or district-level assessment. It will more often relate to a complex thinking process that is the result of applied critical-reasoning skills. (Example: drawing a conclusion, making a recommendation)	What do I want to be sure students are more competent doing when this performance is complete? Do I want them to be able to develop a range of possible solutions to a problem? Will they investigate an issue from outside school, form an opinion, or describe and support a point of view? What complex thinking skill is the core purpose of this performance?
❷ KEY ORGANIZING QUESTION As with the purpose, the question focuses and organizes the entire performance. It combines with the role and the audience to define the context.	What will the students be accessing information about? Do I want to select the issue or question to be accessed, or will the students determine the learning they will pursue? Is the question or issue developmentally appropriate, and can I facilitate obtaining the resources that students will need for the issue? Do the students have any experiential background for this issue? Will the experience be limited to learning from the experiences of others?
❸ ROLE When students take on a role, the point of view of the role adds a dimension not common to most learning. The role introduces the prompt that initiates the entire performance.	Will this role be authentic? Or is it a role-play? For example, students as artists, authors, and investigators are real roles for students. Students as lawyers, policemen, or city council members do not have the same level of authenticity. They would be role-playing, which is pretending to be someone. Will there be more than one role or will students all be in the same role? How will I ensure that students will have a focused point of view to explore? In life outside school, who would answer this question or be concerned with this issue? What would that expert do? Who is the expert? What's the real role?

Performance Designers

Section II

The second section of the performance designer focuses on having students carry out the learning actions of accessing and interpreting necessary content and concepts. The right-hand column of the top section defines the quality criteria, or "look fors," that will be taught, practiced, and assessed. These are the quality criteria of the performance benchmarks.

PERFORMANCE DESIGNER ELEMENT	REFLECTIVE QUESTIONS
D ACCESS AND Accessing actions might require students to interview, locate, or read for information. The importance of student involvement in acquiring information requires a shift from teacher as information provider to teacher as facilitator for information accessing. **E INTERPRET BY...** *(Do what?)* Interpreting actions require students to review what they have collected and decide what it means now that they have it. Students may categorize the information they have, compare it with what they already know, and process it in a variety of critical and creative ways.	Where can information be accessed? Are there experts who can be interviewed? What publications will be helpful? Which texts contain related information? Who can we contact on the Internet? What other resources are available? How will students interact with the information they have collected? Will they formulate new questions? Will they begin to be asked to draw conclusions or perhaps make predictions at this point in the performance? Who will students interact with to communicate their initial interpretations of the information? Will they have a peer conference? Will I ask questions or give answers?
F CONTENT/CONCEPTS *(With what?)* This specifies the knowledge or information the students are to learn. The result at the end is only going to be as good as the information the students collect. The resources should go far beyond the text. The teacher should support with additional resources and literature examples.	What do the students need to know? Where will the information come from? What will be significant learning to retain after the performance is over? Why is it important for students to learn this? Where might they need to use it later? Next year? After they leave school? What connections can they make to other knowledge structures? What are different points of view?
G QUALITY CRITERIA **("Look fors")** *(How well?)* Quality criteria are the specifications for the performance. It is critical that these "look fors" be observable and measurable and that they represent high-quality performances. The quality criteria stated in the third column will integrate the learning actions in the left-hand column with the content/concept to be learned in the center column.	What would an expert interviewer or artist do? What would be observable in the performance of a quality questioner or researcher? How would I know one if I saw one? Do the criteria match the learning action that has been selected, and do they describe a logical and relevant application of the content/concept that is to be learned?

Section III

The third main section on the performance designer is organized similarly but focuses on the producing competence in the Learning Actions Wheel. The middle column of this section allows the teacher to describe or specify the nature of the product or performance the students are to generate. The right-hand column describes the quality criteria, or "look fors," that pertain to that product or production.

PERFORMANCE DESIGNER ELEMENT	REFLECTIVE QUESTIONS
❶ PRODUCE BY... *(In order to...* Producing actions ask students to synthesize their learning, to bring what has been learned together into a cohesive whole that has relevance. Students might design, build, develop, create, construct, or illustrate.	How will students bring what they have learned together? What actions will lead to a product and keep the students in the role? Are there stages to the producing action, such as design and develop or draft and write? What are the essential actions that will lead to a product?
❶ PRODUCT/PERFORMANCE *...do what?)* This describes the product, service, or production that the student will address. It should be something that will benefit the authentic audience.	What would an expert create? How does this product relate to the required or identified knowledge base? How does this product incorporate the required skills? What impact should the product have on the audience?
❶ QUALITY CRITERIA *("Look fors")* *(How well?)* Quality criteria describe the learning actions as they occur in conjunction with the development of the product, service, or production. It is critical that the criteria be observable and measurable, and that it represent quality.	What would an excellent product look like? How could it be described? Will the product or production indicate the designing and developing that were used? How can it be precisely described in relationship to the learning actions? What are the essential actions the student will perform that relate to the producing verbs?

Section IV

The last section of the performance designer relates to the disseminating learning actions. It describes the sharing of the product. The middle column of this section clearly denotes the *audience* and the *setting*, or *context*, in which the performance will occur. These factors are critical in determining the realistic impact of the student's learning. The right-hand column will describe the quality criteria for this portion of the performance designer by combining the disseminating action in relationship with the authentic audience. It defines the purpose for the learning.

PERFORMANCE DESIGNER ELEMENT	REFLECTIVE QUESTIONS
🅚 DISSEMINATE BY... Learning actions at this stage of the role performance or demonstration have the learners presenting their products, services, or productions. The form of the presentation will vary depending on the original purpose. The learner might disseminate by explaining, teaching, or dancing.	What will be the most efficient and effective form of communicating this new product? Will students choose to broadcast or publish or teach? How does this form best relate to the product and the purpose? How does this delivery relate to the role?
🅛 AUDIENCE/SETTING *(To/for whom? Where?)* The audience will be the recipient of the learners' product or production. The degree of authenticity will be reflected in the composition of the audience. The setting for this performance could be related to the original issue being investigated as well as the purpose for this investigation, or the natural location of the recipient.	Who will benefit from the students' learning? Who can use this recommendation or this finding? Is it another learner? Someone at another grade level? Is it a team of engineers at General Motors? Or young patients in a dentist's office? Where is the audience?
🅜 QUALITY CRITERIA *("Look fors")* *(How well?)* The criteria describe the specification for delivering the product, service, or performance. They are observable and represent quality.	What does a quality presentation look like? What are the essential elements that clearly define a quality presentation? How do the criteria connect the disseminating actions with the learner and the audience?

The performance designer gives teachers a very useful tool for continuously defining learning in terms of a realistic role that students must either individually or collectively take on and accomplish. The performance designer also continuously engages students in the range of learning actions that successful people engage in after they graduate from school, but it does so in the safe environment of school under the careful guidance of the teacher. Learners will demonstrate each role performance according to their developmental level of growth. Continuous involvement and experience with learning actions and quality criteria will result in demonstrated student improvement and continuous upleveling of quality criteria that will fully prepare students for any performance benchmark they are asked to demonstrate.

EXAMPLES OF LEARNING ACTIONS

ACCESS:
Investigate
Gather
Interview
Research
Listen
Observe
Collect
Search
Inquire
Survey
View
Discover
Read
Explore
Examine

INTERPRET:
Analyze
Explain
Paraphrase
Rephrase
Clarify
Compare
Contrast
Summarize
Integrate
Evaluate
Translate
Prioritize
Synthesize
Sort
Classify

PRODUCE:
Create
Design
Develop
Draw
Write
Lay out
Build
Draft
Invent
Erect
Sketch
Assemble
Compose
Illustrate
Generate

DISSEMINATE:
Publish
Perform
Teach
Present
Transmit
Display
Explain
Broadcast
Act
Advertise
Discuss
Send
Sing
Dance
Telecast

EVALUATE:
Review
Reflect
Assess
Revisit
Compare
Conclude
Generalize
Prove
Question
Refute
Support
Verify
Test
Realign
Judge

Performance Designers

SAMPLE PERFORMANCE DESIGNER FOR GRADE 3

PURPOSE:
To make connections

SCIENCE:
PHYSICAL SCIENCE–MATTER
(SEE PAGE 29)

KEY ORGANIZING QUESTION:
How and why are common measuring instruments used?

ROLE: *(You are ...)*
An information collector

(Who is expected to ...)

COMPETENCE (Do what?)	CONTENT/CONCEPTS (With what?)	QUALITY CRITERIA ("Look fors")
Access and interpret by ... collecting analyzing	information by using various measuring instruments on a variety of objects in your home. how the instruments helped you get different kinds of data.	• Clearly state your purpose. • Identify all possibilities. • Select your tools and objects to measure. • Record needed data. • Identify differences between results collected. • Explain the differences.

COMPETENCE (In order to ...)	PRODUCT/PERFORMANCE (... do what?)	QUALITY CRITERIA ("Look fors")
Produce by ... creating	a detailed chart of objects measured, tools used for measuring, locations, measurements collected, and differences noted.	• Identify a plan. • Identify various categories. • Develop a form. • Represent the data clearly and accurately. • Review and adjust as needed.

COMPETENCE	AUDIENCE / SETTING (To/for whom? Where?)	QUALITY CRITERIA ("Look fors")
Disseminate by ... presenting explaining	the various data you collected to a partner at school. how the tools helped you collect information.	• Organize your ideas. • Practice alone. • Share your information with a partner. • Include relevant information. • Supply necessary details. • Ask for reactions or suggestions.

SAMPLE PERFORMANCE DESIGNER FOR GRADE 5

PURPOSE:
To determine cause and effect and to make a recommendation

SCIENCE:
LIFE SCIENCE–ECOSYSTEMS
(SEE PAGE 24)

KEY ORGANIZING QUESTION:
Why are some species of plants and animals becoming extinct?

ROLE: *(You are ...)*
A scientific investigator

(Who is expected to ...)

COMPETENCE (Do what?)	CONTENT/CONCEPTS (With what?)	QUALITY CRITERIA ("Look fors")

Access and interpret by ...
   gathering
 	
   analyzing
	information on a topic related to a cause of animal extinction.
	
	 the information to determine cause and effect.
		• Identify your task.
		 • Decide where to get materials and information.
		 • Collect information from a variety of sources.
		 • Identify major points.
		 • Use two different techniques to establish cause and effect.
		 • Select most appropriate information.

COMPETENCE (In order to ...)	PRODUCT/PERFORMANCE (... do what?)	QUALITY CRITERIA ("Look fors")

Produce by ...
   designing
   and
   developing
	a poster for your local environmental agency.	• Identify your audience.
		 • Select the appropriate information.
		 • Include accurate details.
		 • Create a draft.
		 • Check for impact and accuracy.
		 • Create a finished product.

COMPETENCE	AUDIENCE / SETTING (To/for whom? Where?)	QUALITY CRITERIA ("Look fors")

Disseminate by ...
   presenting
   and
   explaining
	your poster to students in another class and representatives from the local environmental agency or center.	• Select appropriate method for presenting.
		 • Organize your idea and messages.
		 • Include relevant information.
		 • Deliver copy to an environmental center or agency.

Performance Designers 81

SAMPLE PERFORMANCE DESIGNER FOR GRADE 8

PURPOSE:
To compare and contrast, and publish

SCIENCE:
LIFE SCIENCE–ECOSYSTEMS
(SEE PAGE 25)

KEY ORGANIZING QUESTION:
What is the relationship of an organ to its organism and an organism to its environment?

ROLE: *(You are ...)*
A scientific investigator

(Who is expected to ...)

COMPETENCE (Do what?)	CONTENT/CONCEPTS (With what?)	QUALITY CRITERIA ("Look fors")
Access and interpret by ... collecting	information on the habits, needs, and adaptability of the water turtle, the box turtle, and the aquatic frog in a planned environmental center.	• Identify your purpose. • Decide on resources. • Select the needed information on each of the creatures. • Select important details.
summarizing	the gathered information.	• Condense the information.

COMPETENCE (In order to ...)	PRODUCT/PERFORMANCE (... do what?)	QUALITY CRITERIA ("Look fors")
Produce by ... designing and developing	a graphic chart that clearly depicts the unique features, characteristics, and needs of each of these creatures along with their adaptability and survival.	• State your purpose. • Identify key categories. • Create a representation. • Review and edit for accuracy and needed information. • Create the final product. • Extend ideas to the fullest.

COMPETENCE	AUDIENCE / SETTING (To/for whom? Where?)	QUALITY CRITERIA ("Look fors")
Disseminate by ... presenting	your graphic chart to another team in your class.	• Organize your ideas. • Deliver your ideas to team members. • Review reactions.
submitting	the finished product.	• Adjust chart as needed. • Deliver final product.

SAMPLE PERFORMANCE DESIGNER FOR GRADE 12

PURPOSE:
To solve a problem and make recommendations

SCIENCE:
LIFE SCIENCE–LIVING THINGS
(SEE PAGE 20)

KEY ORGANIZING QUESTION:
How concerned should we be about microbes causing food spoiling or poisoning?

ROLE: *(You are ...)*
A health investigator

(Who is expected to ...)

COMPETENCE (Do what?)	CONTENT/CONCEPTS (With what?)	QUALITY CRITERIA ("Look fors")
Access and interpret by ... researching translating	information on preservation of food, food spoilage, and food poisoning. the key information into recommended procedures for food handlers and servers.	• Identify your purpose. • Utilize a variety of resources. • Organize your information. • Document references. • Reword key ideas and information for application by food servers.

COMPETENCE (In order to ...)	PRODUCT/PERFORMANCE (... do what?)	QUALITY CRITERIA ("Look fors")
Produce by ... designing and developing	an informational guide on safe food handling at family or public outings.	• Identify purpose. • Brainstorm key sections. • Include results from microbiological tests. • Consider most often asked questions. • Create a representation. • Review and edit as necessary. • Create final product.

COMPETENCE	AUDIENCE / SETTING (To/for whom? Where?)	QUALITY CRITERIA ("Look fors")
Disseminate by ... publishing distributing	the final guide on safe food handling. the guide to the home economics department and families of students or through the local health department.	• Organize the layout. • Consider visual appeal. • Duplicate and compile. • Deliver to intended audience.

Performance Designers

SAMPLE PERFORMANCE DESIGNER FOR GRADE 3

PURPOSE:
To draw a conclusion

SCIENCE:
EARTH SCIENCE–METEOROLOGY
(SEE PAGE 40)

KEY ORGANIZING QUESTION:
How does temperature affect the simple responses of living things?

ROLE: *(You are ...)*
A health investigator

(Who is expected to ...)

COMPETENCE (Do what?)	CONTENT/CONCEPTS (With what?)	QUALITY CRITERIA ("Look fors")
Access and interpret by ... gathering clarifying	information on body temperatures and thermometers. the most important aspects of the gathered information.	• Identify your topic. • Decide where to get the information. • Collect information from several sources. • Select the important points. • State this information clearly to another person.

COMPETENCE (In order to ...)	PRODUCT/PERFORMANCE (... do what?)	QUALITY CRITERIA ("Look fors")
Produce by ... planning and conducting	an experiment to see if physical exercise will cause change in body temperature.	• Identify your purpose. • Create a step-by-step plan for your experiment. • Gather necessary materials and resources. • Execute your plan. • Record needed data.

COMPETENCE	AUDIENCE / SETTING (To/for whom? Where?)	QUALITY CRITERIA ("Look fors")
Disseminate by ... publishing delivering	your findings in a guide for joggers. your guides to members of a jogging club.	• Select necessary information and data. • Format the information. • Organize the layout. • Check for accuracy. • Produce the guide. • Deliver to users.

SAMPLE PERFORMANCE DESIGNER FOR GRADE 5

PURPOSE:
To transfer information and make predictions

SCIENCE:
PHYSICAL SCIENCE—ENERGY
(SEE PAGE 36)

KEY ORGANIZING QUESTION:
How effective are various materials as insulators or conductors?

ROLE: *(You are ...)*
A scientific investigator

(Who is expected to ...)

COMPETENCE (Do what?)	CONTENT/CONCEPTS (With what?)	QUALITY CRITERIA ("Look fors")
Access and interpret by ... collecting summarizing	information on how well different materials conduct heat by conducting experiments. the observed results of the experiment.	• Identify possible targets. • Identify tools. • Review procedures. • Conduct experiments. • Record observation accurately. • Select the main ideas. • Provide needed details. • Condense the information.

COMPETENCE (In order to ...)	PRODUCT/PERFORMANCE (... do what?)	QUALITY CRITERIA ("Look fors")
Produce by ... drafting predicting	an outline of a house including doors, windows, roof, and chimney. areas of heat lost in this house based on information obtained from the experiments.	• Identify your topic. • Include necessary items. • Highlight areas you predict will result in heat loss. • Clearly labels areas of heat loss and cause. • Offer possible solutions on the drawing.

COMPETENCE	AUDIENCE / SETTING (To/for whom? Where?)	QUALITY CRITERIA ("Look fors")
Disseminate by ... publishing presenting	your final drawing with predictions and possible solutions. your drawing and explaining it to members of another team and to your parents.	• Present a clean, neat drawing to an audience. • Organize your ideas. • Deliver your message. • Ask for suggestions or new ideas.

Performance Designers 85

SAMPLE PERFORMANCE DESIGNER FOR GRADE 8

PURPOSE:
To make predictions, assess, and observe

SCIENCE:
EARTH SCIENCE–METEOROLOGY
(SEE PAGE 42)

KEY ORGANIZING QUESTION:
How can various tools help predict changes in weather?

ROLE: *(You are ...)*
A weather forecaster

(Who is expected to ...)

COMPETENCE (Do what?)	CONTENT/CONCEPTS (With what?)	QUALITY CRITERIA ("Look fors")
Access and interpret by ... gathering analyzing	information on barometers in predicting weather changes and information from observing a self-made barometer and a professional barometer. the information collected from observing the two barometers.	• Clearly state your task and your procedure. • Collect all needed information and materials. • Predict outcomes. • Conduct the experiments. • Record information from observations. • Prioritize information. • Compare information from the two barometers.

COMPETENCE (In order to ...)	PRODUCT/PERFORMANCE (... do what?)	QUALITY CRITERIA ("Look fors")
Produce by ... creating	a presentation for your classmates including visuals, charts, and graphs of your gathered information and the results.	• Identify purpose of presentation. • Review possible directions. • Select most useful idea. • Develop a draft of items. • Review and check for details. • Develop final copy of materials.

COMPETENCE	AUDIENCE / SETTING (To/for whom? Where?)	QUALITY CRITERIA ("Look fors")
Disseminate by ... presenting and explaining	your study, the results, and the visuals to a team of classmates and other interested audiences.	• Organize your ideas. • Organize your materials. • Clearly present major procedures and findings. • Elicit feedback and respond accordingly.

SAMPLE PERFORMANCE DESIGNER FOR GRADE 12

PURPOSE:
To create explanation with supporting examples

SCIENCE:
PHYSICAL SCIENCE–ENERGY
(SEE PAGE 38)

KEY ORGANIZING QUESTION:
How does force affect motion?

ROLE: *(You are ...)*
A motion investigator

(Who is expected to ...)

COMPETENCE (Do what?)	CONTENT/CONCEPTS (With what?)	QUALITY CRITERIA ("Look fors")
Access and interpret by ... researching and analyzing	information on Newton's law of motion and related examples from daily life.	• Design and organize questions. • Utilize a variety of resources. • Organize collected information. • Identify big ideas. • Compare and contrast the key ideas. • Prioritize the information as it relates to your key question.

COMPETENCE (In order to ...)	PRODUCT/PERFORMANCE (... do what?)	QUALITY CRITERIA ("Look fors")
Produce by ... drafting and writing	an article for an elementary magazine that explains the law of motion as it relates to various daily examples.	• Clearly state your topic. • Create a preliminary copy of ideas. • Review your work. • Edit and change as needed for clarity. • Tie conclusion to original idea, purpose. • Prepare final draft and copy.

COMPETENCE	AUDIENCE / SETTING (To/for whom? Where?)	QUALITY CRITERIA ("Look fors")
Disseminate by ... presenting submitting	to fellow students. to science magazine for publication.	• Outline your ideas. • Rehearse your presentation. • Deliver to students. • Review and edit based on their reactions and suggestions. • Send final copy to several sources for publishing.

APPENDIX: BLANK TEMPLATES

Appendix: Blank Templates

PERFORMANCE DESIGNER

PURPOSE:

KEY ORGANIZING QUESTION:

ROLE: *(You are ...)*

(Who is expected to ...)

COMPETENCE *(Do what?)*	CONTENT/CONCEPTS *(With what?)*	QUALITY CRITERIA *("Look fors")*
Access and interpret by ...		

COMPETENCE *(In order to ...)*	PRODUCT/PERFORMANCE *... do what?)*	QUALITY CRITERIA *("Look fors")*
Produce by ...		

COMPETENCE	AUDIENCE / SETTING *(To/for whom? Where?)*	QUALITY CRITERIA *("Look fors")*
Disseminate by ...		

Burz and Marshall. *Performance-Based Curriculum for Science: From Knowing to Showing.* © 1997 by Corwin Press, Inc.

Appendix: Blank Templates

Science: Performance
Grade ___ Benchmark

CONTENT/CONCEPT STANDARD ___

KEY ORGANIZING QUESTION:

KEY COMPETENCES	KEY CONCEPTS AND CONTENT	PERFORMANCE TASKS
		PERFORMANCE TASK I:
		PERFORMANCE TASK II:

QUALITY CRITERIA:

Burz and Marshall. *Performance-Based Curriculum for Science: From Knowing to Showing.* © 1997 by Corwin Press, Inc.

Appendix: Blank Templates

Technology Connections
_____ : Grade ___

Performance Benchmark

KEY ORGANIZING QUESTION:

ACCESS	PRODUCE	DISSEMINATE
PERFORMANCE TASK I:	**PERFORMANCE TASK I:**	**PERFORMANCE TASK I:**
PERFORMANCE TASK II:	**PERFORMANCE TASK II:**	**PERFORMANCE TASK II:**

Burz and Marshall. *Performance-Based Curriculum for Science: From Knowing to Showing.* © 1997 by Corwin Press, Inc.

BIBLIOGRAPHY

Aldridge, W. G. (1992). *Scope, sequence and coordination of secondary school science: The content core.* Washington, DC: The National Science Teachers Association.

American Association for the Advancement of Science. (1989). *Project 2061: Science for all Americans.* Washington, DC: Author.

American Association for the Advancement of Science. (1993). *Project 2061: Benchmarks for science literacy.* New York: Oxford University.

Brooks, J. G., & Brooks, M. G. (1993). *In search of understanding: The case for constructivist classrooms.* Alexandria, VA: Association for Supervision and Curriculum Development.

California State Department of Education. (1990). *Science framework for California public schools: Kindergarten through grade twelve.* Sacramento, CA: Author.

Caine, R. N., & Caine, G. (1991). *Making connections: Teaching and the human brain.* Alexandria, VA: Association for Supervision and Curriculum Development.

Cheek, D. W., Briggs, R., & Vager, R. (1992). *Science curriculum resource handbook.* Milwood, NY: Kraus International.

Coleman, W. (1993). *Educating Americans for the 21st century: A report of the National Science Board, Commission on Precollege Education in Mathematics, Science and Technology.* Washington, DC: Author.

Florida Science Framework Commission. (1993). *Science for all students.* Tallahassee: Florida Department of Education.

Hacker, M., & Barden, R. (1988). *Living with technology.* Albany, NY: Delmar.

Hazen, R. M., & Trefil, J. (1990). *Science matters: Achieving scientific literacy.* New York: Doubleday.

Kent County Intermediate School District, & Gordon, T. (1992). *Elementary technology teachers' guide.* Grand Rapids, MI: Author.

King, J., & Evans, K. (1991). Can we achieve outcome-based education? *Educational Leadership, 49*(2), 73-75.

LaPointe, A. E., Askew, J. M., & Mead, N. A. (1991). *Learning science: The international assessment of educational progress.* Princeton, NJ: Educational Testing Service.

McFadden, C., & Vager, R. (1993). *Science plus: Technology and society.* Chicago, IL: Holt, Rinehart & Winston.

Michigan Department of Education. (1989, 1992). *Michigan-Ohio technology education consortium.* East Lansing, MI: Michigan State University.

Michigan Department of Education. (1994). *Michigan technology framework.* Lansing, MI: Author.

National Center for Education Statistics. (1992). *The 1990 science report card: NAEP's assessment of fourth, eighth, and twelfth graders.* Washington, DC: U.S. Department of Education.

State of Connecticut Board of Education. (1991). *A guide to curriculum development: Science.* Hartford, CT: Author.

Viadero, D. (1993, March 10). The coherent curriculum. *Education Week,* 41-42.

Virginia Polytechnic Institute and State University. (1992). *Kids and technology: Mission 21.* Albany, NY: Delmar.

The Corwin Press logo—a raven striding across an open book—represents the happy union of courage and learning. We are a professional-level publisher of books and journals for K–12 educators, and we are committed to creating and providing resources that embody these qualities. Corwin's motto is "Success for All Learners."